D1710608

CalcLabs with *Mathematica*®

Nancy R. Blachman
Colin P. Williams
Albert Boggess
David Barrow
Arthur Belmonte
Samia Massoud
Jeffrey Morgan
Maurice Rahe
Michael Stecher
Philip Yasskin

Brooks/Cole Publishing Company

I(T)P An International Thomson Publishing Company

Pacific Grove Albany Bonn Boston Cincinnati Detroit
London Madrid Melbourne Mexico City New York Paris
San Francisco Singapore Tokyo Toronto Washington

Brooks/Cole Publishing Company
A division of International Thomson Publishing Inc.

Printed in the United States of America

10 9 8 7 6 5 4 3 2 1

Library of Congress Cataloging-in-Publication Data
CalcLabs with Mathematica / Nancy Blachman . . . [et al.].
p. cm. — (Brooks/Cole symbolic computation series)
Includes index.
ISBN 0-534-34086-5
1. Calculus–Data processing. 2. Mathematica (Computer file)
I. Blachman, Nancy. II. Series.
QA303.5.D37C343 1996
515′.078—dc20 95-34374
CIP

Sponsoring Editor: *Robert Evans*
Marketing Team: *Adrian Perenon, Nancy Conti*
Marketing Representative: *Ragu Raghavan*
Editorial Associate: *Nancy Conti*
Production Editor: *Marlene Thom*
Manuscript Editor: *Lyn Dupré*
Interior Illustration: *Lisa Torri*
Cover Design: *Vernon T. Boes*
Printing and Binding: *Malloy Lithographing, Inc.*

Contents

Preface

Mathematica is a powerful software tool for mathematical computations and visualization. *Mathematica* is radically changing the way scientists and engineers do mathematics in much the same way that calculators changed the computational landscape in the 1970s. The goal of this manual is to introduce this software to students who are currently taking first-year calculus. In this book, we focus on the use of *Mathematica* as a tool to solve problems that would be difficult to solve by hand. We hope that, by using *Mathematica* to solve problems, students will also improve their understanding of the concepts of calculus.

This manual is written for a calculus course that involves 1 hour of computer laboratory contact per week (with some additional open laboratory hours available outside of class). Many large universities cannot offer more than 1 contact hour per week in a computer laboratory because of budget constraints and the ever increasing enrollments in calculus. For this reason, the goals of this manual are rather modest. The manual uses relatively few *Mathematica* commands in order to keep the amount of syntax to a minimum. Most of the examples and exercises involve little formal programming with *Mathematica*. Indeed, one of the beautiful qualities of *Mathematica* is that much can be done with few commands. As another disclaimer, this manual will not magically elucidate all the concepts of calculus (although we hope it will help).

The manual is divided into two parts. The first part (Chapters 1 though 14) reads like a standard text, introducing the use of *Mathematica* with examples. Exercises are given at the end of each of these chapters. Most of the *Mathematica* commands needed for a first-semester calculus course are contained in Chapters 1, 2, 3, 4, and 7. At a bare minimum, students should learn the following skills from these sections: how to assign variables; how to represent functions, expressions, and equations in *Mathematica*; how to graph functions, expressions and equations; how to solve equations; and how to differentiate and integrate functions and expressions. Section 2.4, which describes some of the options of the `Plot` command, can be skipped at first and picked up later, as needed. Most of the remaining chapters in the first part focus more on using previously defined *Mathematica* commands to solve calculus problems. Some additional *Mathematica* commands are introduced as needed. Chapter 13 contains material on programming with *Mathematica*, as well as descriptions of some of the more sophisticated *Mathematica* commands. The second part of the manual (Chapters 15 through 16) contains student projects and weekly laboratories that can be assigned as time permits. Some of the projects are meant as longer research projects for the more motivated students (e.g., the brightest phase of Venus, and the project on gravitational force).

The exercises and examples vary considerably in length and difficulty. Some of the exercises and examples are routine, and are designed to illustrate *Mathematica* syntax. Others are more involved, and are designed to solve problems or to illustrate ideas that would be difficult using only hand computation. Many of the more complicated examples and exercises are designed to embellish classical problems with real-life considerations. The following example (presented in Chapter 6) illustrates this aspect. Consider the problem of minimizing the surface area of a cylindrical can of fixed volume. Most calculus books in print over the past several generations contain this problem as an example or an exercise. After this problem has been translated into mathematics, its solution is straightforward and does not require the use of a computer. However, the computer allows the student to consider a more real-life version of this problem that seeks to minimize the cost of constructing this can where the cost of the seam used to attach the top, bottom, and sides is considered, along with the cost of the material. Minimizing this cost function requires the use of the computer for computations and graphics. Students should be expected to do the classical version of this problem by hand, and then they can be assigned the more real-life version of this problem in the computer laboratory.

As a word of warning to students, *Mathematica* alone will not solve calculus problems. You must still do the thinking required for problem set up. *Mathematica* can then be used to help with the computations and graphics that are necessary to obtain final answers. With this in mind, students should read the text of each chapter and set up any assigned problems before coming to their computer laboratory to make the best use of time spent in front of the computer terminal.

As a final note, all the *Mathematica* syntax and examples presented in this manual use *Mathematica* Version 2.2. Most, but not all, of the syntax presented in this manual is valid for earlier versions of *Mathematica*. This manual assumes that you are using the Notebook Front End version of *Mathematica*, which runs on *X/Motif, Microsoft Windows*, *NeXT*, and *Macintosh*); however, references to platform-specific features of *Mathematica* (e.g., menus) have been avoided so as not to tie this manual to any specific computer platform. Other users can still use this manual, but they will have to make some modifications in syntax. This manual was written with *Mathematica* Version 2.2, and then exported as a LaTeX file to ensure that no errors were made in transcribing *Mathematica* output into the text.

How to Read This Book

This book contains *Mathematica* input and output. The input is typeset in **`a bold typewriter face, like this,`** and the output is in `a light typewriter face, like this.` *Mathematica* keywords in the text are also set in the `upright` `typewriter type`. When we want to show you the result of executing a particular piece of *Mathematica* code, we will display it like this:

```
In[1]:=  Integrate[ 1/(1+x^3), x ]

Out[1]=          -1 + 2 x
         ArcTan[--------]                                  2
                  Sqrt[3]      Log[1 + x]    Log[1 - x + x ]
         ---------------- + ---------- - ---------------
              Sqrt[3]              3                6
```

Sometimes, we will show you a piece of "generic" *Mathematica* code, as a pattern to follow when writing your own *Mathematica* input. Here is an example of such a pattern:

```
Plot[function, {variable, start, end}]
```

The words and symbols in typewriter type must be typed exactly as shown, but the words in *slant roman type, like this,* must be replaced by actual *Mathematica* expressions. One way of making replacements in the previous pattern would be

```
Plot[Sin[x], {x, 0, 2Pi}]
```

Here, `Sin[x]` is the *function,* `x` is the *variable,* `0` is the *start* value, and `2Pi` is the *end* value. When the result of executing a command includes a picture (the typical case in this book), it will be typeset like this: `Plot[Sin[x], {x, 0, 2Pi}]`.

Software

All the input statements in this book are available through ftp from the host `brookscole.com`. There are files containing the inputs from the chapters. Files with names that terminate in "`.ma`" are intended to be opened with the *Mathematica* Notebook Front End. The inputs from Chapters 15 and 16 may be found in the directories `chap15` and `chap16`, respectively.

```
chap1.ma   chap5.ma   chap9.ma    chap13.ma
chap2.ma   chap6.ma   chap10.ma   chap14.ma
chap3.ma   chap7.ma   chap11.ma   chap15
chap4.ma   chap8.ma   chap12.ma   chap16
```

Start up the Notebook Front End, and then open up a file by selecting the menu item *Open* in the *File* menu. You can generate or regenerate a graphic by putting your cursor in an input line or cell and (1) pressing the ⟨ENTER⟩ key, (2) pressing the ⟨SHIFT⟩ and ⟨RETURN⟩ keys simultaneously, or (3) selecting the menu item *Evaluate Selection* from the *Action* menu. Files with names that terminate in "`.m`" are intended to be used on systems running *Mathematica* without the Notebook Front End, such as MS-DOS or Unix. These files are in the directory or folder named "`dos-unix`."

chap1.m	chap5.m	chap9.m	chap13.m
chap2.m	chap6.m	chap10.m	chap14.m
chap3.m	chap7.m	chap11.m	chap15
chap4.m	chap8.m	chap12.m	chap16

Start up *Mathematica*, and then read a file by invoking the command "`<<` *filename*," where *filename* is one of the files in the directory called "`dos-unix`." You may need to resize the graphics to make them resemble the images in the book. If your computer has a small amount of RAM, you may not be able to run all the inputs on this disk during a single *Mathematica* session. When an input does not produce the results shown in the book, we recommend that you start a new *Mathematica* session and reexecute the input.

The History of the Book

Albert Boggess, David Barrow, Arthur Belmonte, Samia Massoud, Jeffery Morgan, Maurice Rahe, Michael Stecher, and Philip Yasskin teach a calculus laboratory using the Maple computer algebra system at Texas A & M University. They recently published their course notes in the book *CalcLabs with Maple V*. Because of the popularity of this book and of *Mathematica*, Brooks/Cole decided to publish a version of the book based on *Mathematica*. Gary Ostedt, a math editor at Brooks/Cole, approached Nancy Blachman in the spring of 1995 about this idea, because she has taught courses and workshops on *Mathematica* and written several popular books on*Mathematica* (*Mathematica: A Practical Approach*, *The Mathematica Quick Reference*, and *The Mathematica Graphics Guidebook*) and a comprehensive guide on Maple (*The Maple V Quick Reference*). Bob Evans and others at Brooks/Cole wanted to have *CalcLabs with Mathematica* published and available by the fall of 1995. To meet the tight deadline, Nancy invited Colin Williams, an expert on *Mathematica*, to co-author this translation.

How this Book was Produced

Mathematica commands were tested on several platforms for this guide—including NeXT workstations, Apple Macintoshes, and IBM-compatible personal computers—using *Mathematica*, Version 2.2. The book was typeset with the LaTeX document-preparation system. Most of the PostScript illustrations were produced by *Mathematica*, and were included in the book with Radical Eye Software's epsf macros.

The dvips program by Radical Eye Software rendered a PostScript file from the TEX output.

Acknowledgments

Nancy and Colin — We are grateful to all who gave us suggestions for this guide. We thank Lyn Dupré, Xah Lee, and several anonymous reviewers for offering us constructive feedback. We appreciate Don DeLand and Cameron Smith's assistance with TEX. It has been a delight to work with the people at Brooks/Cole, including Bob Evans, our editor, Nancy Champlin Conti, Bob's editorial assistant, and Marlene Thom, our production editor.

Nancy — I thank my parents, Nelson and Anne Blachman, for believing in me and encouraging me to pursue my desires.

I thank Gary Ostedt for suggesting that I translate the book *CalcLabs with Maple V* to *Mathematica*.

I also thank Alesia Bland for handling Variable Symbols business while I was working on this guide.

Colin — Most of all, I thank my wife, Patricia, for her patience, support, and encouragement.

Request for Feedback

We hope you find the *CalcLabs with Mathematica* useful. If you come across any errors or omissions, or if you have suggestions for how this guide can be improved, please tell us.

Nancy Blachman
Variable Symbols, Inc.
Email: `nb@cs.stanford.edu`
Fax: 510-652-8461

Colin P. Williams
415-728-2118
Email: `CPWilli118@aol.com`

Common *Mathematica* Commands

Here is a list of common *Mathematica* commands. Be aware that all *Mathematica* commands begin with an uppercase letter. You can obtain help on the syntax of any *Mathematica* command by typing `?`*command*. For example, to get help with the `Solve` command, type `?Solve`.

Assignments

`a = 1.53`
: Assigns the value 1.53 to the value `a`.

`a = Pi r^2`
: Assigns the name `a` to the given formula.

`Clear[`*a*`]`
: Clears any value previously given to *a*.

`f[x_] := x^2 + 5`
: Defines the function $f(x) = x^2 + 5$.

Numerical and Algebraic Manipulation

`N[`*expr*`]`
: Gives a numeral approximation for *expr*.

`Expand[`*expr*`)]`
: Expands out products and powers in the expression *expr*, e.g., expanding $(x+1)^2$ returns $1 + 2x + x^2$.

`Simplify[`*expr*`]`
: Simplifies the expression *expr* algebraically.

`Factor[`*expr*`]`
: Factors the polynomial expression *expr*.

`Apart[`*expr*`]`
: Rewrites the rational expression *expr* as a sum of terms with minimal denominators. In other words, it returns the partial fractions of a rational expression.

Solving Equations

`Solve[`*eqn*`,` *var*`]`

Solves the equation *eqn* for the variable *var*.

`Solve[{`eqn_1`,` eqn_2`, ...,` eqn_n`}, {`var_1`,` var_2`, ...,` var_m`}]`

Solves the system of *n* equations eq_1, eq_2, ..., eq_n exactly in terms of the *m* variables var_1, var_2, ... var_m.

`FindRoot[`*eqn*`, {`x`,` x_0`}]`

Finds a numerical solution to the equation *eqn* given the starting value $x = x_0$.

`NSolve[`*eqn*`,` *x*`]`

Attempts to find a numerical solution to the equation *eqn* for the variable x.

`NSolve[{`eqn_1`,` eqn_2`, ..., eqn`$_n$`}, {`x_1`,` x_2`, ...,` x_m`}]`

Attempts to find a numerical solution to the n equations eqn_1, eqn_2, eqn_n for the m variables $x_1, x_2, \ldots, x_m$.

expr `/.` *var* `->` *value*

After evaluating *expr*, replaces the variable or expression *var* with *value*, i.e., substitutes *value* in place of each occurrence of *var* in *expr*. For example, the input `3x /. x -> a` returns `3a`.

Mathematica Constants and Functions

`E`

The exact value of the exponential constant e, which is the base for natural logarithms.

`Exp[x]`

The exponential function e^x.

`Infinity`

The symbol that represents ∞, a positive infinite quantity.

`Pi`

The exact constant π.

Graphics

`Plot[`*expr*`, {`x`,` $x_{\min}$`,` $x_{\max}$`}]`

Plots *expr* over the interval $x_{\min} \leq x \leq x_{\max}$.

`Plot[`*expr*`, {`x`,` $x_{\min}$`,` $x_{\max}$`}, PlotRange -> {`$y_{\min}$`,` $y_{\max}$`}]`

Plots *expr* over the interval $x_{\min} \leq x \leq x_{\max}$ restricting the displayed values of y in the range $y_{\min} \leq y \leq y_{\max}$.

`Plot[{`$expr_1$`,` $expr_2$`}, {`x`,` $x_{\min}$`,` $x_{\max}$`}]`

Plots the graphs of $expr_1$ and $expr_2$ over the interval $x_{\min} \leq x \leq x_{\max}$ on the same coordinate axis.

`Plot3D[`*expr*`, {`*x*`, `*a*`, `*b*`}, {`*y*`, `*c*`, `*d*`}]`
: Gives a three-dimensional plot of the expression *expr*.

`Needs["Graphics`ImplicitPlot`"]`
: Loads the `ImplicitPlot`` package, which you need to do before calling `ImplicitPlot`. The function `ImplicitPlot` plots of points that satisfy an equation.

`ImplicitPlot[`*eqn*`, {`*x*`, `*a*`, `*b*`}, {`*y*`, `*c*`, `*d*`}]`
: Draws a graph of a set of points that satisfy the equation *eqn*.

`ParametricPlot[{`*f[t]*`, `*g[t]*`}, {`*t*`, `$t_{\min}$`, `$t_{\max}$`}]`
: Plots the graph of the parametric equations $x = f(t)$ and $y = g(t)$ for $t_{\min} \leq t < t_{\max}$.

`ListPlot[{`y_1`, `y_2`, ..., `y_n`}]`
: Plots a list the n values $y_1, y_2, \ldots, y_n$ where the x coordinates for those points are taken to be $1, 2, \ldots n$.

`ListPlot[{{`x_1`, `y_1`}, {`x_2`, `y_2`}, ..., {`x_n`, `y_n`}}]`
: Plots a list of the n values $[x_1, y_1], [x_2, y_2], \ldots [x_n, y_n]$.

`ListPlot[`*data*`, PlotJoined -> True]`
: Plots a list of values with connecting line segments. The data may be a list of y values or pairs of x and y values.

Calculus-Related Commands

`Limit[`*expr*`, `*x*` -> `x_0`]`
: Evaluates the limit of *expr* as x approaches x_0.

`Limit[`*expr*`, `*x*` -> `x_0`, Direction -> 1]`
: Evaluates the limit of *expr* as x approaches x_0 from the left, i.e., increasing towards x_0.

`Limit[`*expr*`, `*x*` -> `x_0`, Direction -> -1]`
: Evaluates the limit of *expr* as x approaches x_0 from the right, i.e., decreasing towards x_0.

`D[`*expr*`, `*x*`]`
: Differentiates the expression *expr* with respect to x.

`D[`*expr*`, {`*x*`, `*n*`}]`
: Gives the nth partial derivative of *expr* with respect to x.

`D[`*expr*`, `*x*`, `*y*`]`
: Differentiates the expression *expr* with respect to the variables x and y.

`Integrate[`*expr*`, `*x*`]`
: Returns the value of the integral of *expr* with respect to x.

`Integrate[`*expr*`, {`*x*`, `$x_{\min}$`, `$x_{\max}$`}]`
Returns the value of the definite integral of *expr* over the interval $x_{\min} \leq x \leq x_{\max}$.

`Sum[`*expr*`, {`*i*`, `*m*`, `*n*`}]`
Returns the sum of the expression *expr* for i going from m to n, with a step size of 1.

`Sum[`*expr*`, {`*i*`, `*m*`, `*n*`, `*d*`}]`
Returns the sum of the expression *expr* for i going from m to n, with a step size of d.

`Series[`*expr*`, {`*x*`, `x_0`, `*n*`}]`
Returns the power series expansion for the expression *expr* at the point $x = x_0$ to the order $(x - x_0)^n$.

`Normal[`*series*`]`
Converts a power series expansion to a polynomial by eliminating the error term.

`DSolve[`*eqn*`, `*y[x]*`, `*x*`]`
Solves a differential equation for *y[x]* with the independent variable *x*.

`DSolve[{`*eqn*`, `*init*`}, `*y[x]*`, `*x*`]`
Solves a differential equation with initial or boundary conditions for *y[x]* with the independent variable *x*.

`NDSolve[{`*eqn*`, `*init*`}, `*y[x]*`, {`*x*`, `$x_{\min}$`, `$x_{\max}$`}]`
Finds a numerical solution to the differential equation *eqn* with initial or boundary conditions *init* for *y[x]* with the independent variable *x* in the range $[x_{\min}, x_{\max}]$.

Mathematica-Specific Commands

`%`
Returns the previous result calculated by *Mathematica*.

`%%`
Returns the second to last result calculated by *Mathematica*.

`Table[`*expr*`, {`*i*`, `$i_{\min}$`, `$i_{\max}$`, `i_{step}`}]`
Returns a list of values of *expr* with $i = i_{\min}$ to $i = i_{\max}$ using steps i_{step}.

`Needs["`*context*`` ` ``"]`
Loads a file or package specified by `ContextToFilename["`*context*`` ` ``"]`.

1 Getting Started

This chapter introduces some of the basic *Mathematica* commands and the syntax involved with assigning variables, creating plots, and calculating limits.

1.1 *Mathematica* as a Calculator

You enter a *Mathematica* command by typing it on an input line and then pressing the ⟨ENTER⟩ key, or holding down the ⟨SHIFT⟩ key and pressing the ⟨RETURN⟩ key. Try entering

```
In[1]:=  2 + 5

Out[1]=  7
```

(Do not type the *In[1]* input label, since this is provided by the computer.) *Mathematica*'s output, 7, is displayed to the right of the output label *Out[1]*.

If you enter a command incorrectly, then click the mouse on the line that contains the command, edit it, and then re-execute it by pressing ⟨ENTER⟩, or holding down the ⟨SHIFT⟩ key and pressing the ⟨RETURN⟩ key.

Mathematica can do arithmetic. The standard arithmetic operations are

+	addition
-	subtraction
*	multiplication
/	division
^	exponentiation

The standard order of operations is exponentiation before multiplication and division and then addition and subtraction. To be safe, use parentheses to be sure that the operations are performed in the desired order. For example, `(3+4)/7` is not the same as `3+4/7`. Be sure to use parentheses ( ) rather than brackets [ ] or braces { }, which have other meanings in *Mathematica*.

Mathematica is equipped with a large number of standard mathematical functions including:

the square root function	`Sqrt`
the absolute value function	`Abs`
the natural exponential	`Exp`
the natural logarithm	`Log`
the trig functions	`Sin, Cos, Tan, Sec, Csc, Cot`

the inverse trig functions ArcSin, ArcCos, ArcTan, ArcSec, ArcCsc, ArcCot

It is important to understand that *Mathematica* distinguishes between *exact numbers*, such as 2, 1/3, $\sqrt{2}$, and π, and *floating-point numbers*, such as 2.0, 0.333333, 1.414, and 3.14. The number 1/3 is an expression that represents the exact value of one-third, whereas 0.333333 is a floating-point approximation of 1/3.

If all numbers are entered as integers, then *Mathematica* returns an exact answer. For example, enter

```
In[2]:=  (1 + 3)/6

Out[2]=  2
         -
         3
```

Mathematica returns 2/3 rather than a decimal approximation such as 0.666667. To get the numerical or decimal approximation, use the N command. For example, try

```
In[3]:=  N[22/79 + 34/23]

Out[3]=  1.75674
```

Here, **N** evaluates the expression as a floating-point decimal number. An alternative way to obtain a decimal answer is to enter one of the input numbers in an expression as a decimal. For example, if you type

```
In[4]:=  22./79 + 34/23

Out[4]=  1.75674
```

Mathematica returns the answer in decimal form.

1.2 Assigning Variables

You can refer to a previous result by using one or more percent signs. % refers to the last result, %% refers to the second to last result, %%% refers to the third to last result, and so forth. For example, the calculation of $2^6 + 1$ can be performed in two steps. The result of the first operation is passed to the second part as %.

```
In[5]:=  2^6

Out[5]=  64

In[6]:=  %+1

Out[6]=  65
```

It is also possible to reference a particular result with %*n*, where *n* is the number of the output line. *Note*: If you restart *Mathematica*, you cannot refer to results with the numbers assigned during an earlier session.

If you assign a name to a result, you can later reference that result with the assigned name. For example, you can assign the name a to the value of 22./79 + 34/23 by typing

```
In[7]:=  a = 22./79 + 34/23

Out[7]=  1.75674
```

Mathematica commands of this form are called *assignment statements* and the = sign indicates that the symbol on the left is assigned to the value of the expression on the right. Now, you can recall the number 1.75674 by typing a.

For example, to compute $(1.75674)^2$, type

```
In[8]:=  a^2

Out[8]=  3.08614
```

To compute $\frac{1}{1.756741883}$, enter

```
In[9]:=  1/a

Out[9]=  0.569236
```

To compute $\sqrt{1.756741883}$, enter

```
In[10]:=  Sqrt[a]

Out[10]=  1.32542
```

Names can be as long as you like. We advise using names that are descriptive. For example, you can name the retail price and wholesale cost of an item by using the assignment statements

```
In[11]:=  price = 4.95

Out[11]=  4.95

In[12]:=  cost = 2.80

Out[12]=  2.80
```

The value of the profit is then given by

```
In[13]:=  profit = price - cost

Out[13]=  2.15
```

Symbols or names are case sensitive. The name `profit` is different from the name `Profit`. All names built into *Mathematica* begin with an uppercase character or a $ sign. It is a good idea to define names that start with a lowercase letter so that your names can be distinguished easily from commands built into *Mathematica*.

A variable or name keeps its value until it is assigned a new value or until it is cleared. You can clear the value assigned to the variable `cost` by calling the function `Clear`.

```
In[14]:=  Clear[cost]
```

Now the variable `cost` has no value assigned to it.

As another example, enter an expression that describes the area of a circle of radius r.

```
In[15]:=  area = Pi r^2
Out[15]=      2
          Pi r
```

Note that π is entered with an uppercase P. To evaluate this area when $r = 5$, enter `r = 5`. If instead you enter `r=5;`, *Mathematica* still makes the assignment but no longer displays the result. Operations followed by a semicolon are performed, but the output is not printed. In either case, the value $r = 5$ is automatically substituted into `area`.

```
In[16]:=  r = 5;
          area

Out[16]=  25 Pi
```

The command N gives us a numerical approximation to the area of a circle with radius 5.

```
In[17]:=  N[area]

Out[17]=  78.5398
```

1.3 Algebra Commands

We have seen how to manipulate numbers and how to assign names to them. *Mathematica* can also manipulate algebraic expressions involving names or variables.

For example, to expand or multiply out the expression $(3x-2)^2(x^3+2x)$, type

```
In[18]:=  Expand[(3x - 2)^2 (x^3 + 2x)]

Out[18]=              2         3         4       5
          8 x - 24 x  + 22 x  - 12 x  + 9 x
```

It is easy to make a typographical error when typing a complicated expression, such as $(3x-2)^2(x^3+2x)$. To prevent such errors from affecting a *Mathematica* command (such as Expand), first type the expression without the command and execute it. Then, enter the command that manipulates the expression and execute it. For example, the preceding expansion can be performed as follows. First, type

```
In[19]:=  (3x - 2)^2 (x^3 + 2x)

Out[19]=                2            3
          (-2 + 3 x)  (2 x + x )
```

Examine *Mathematica*'s output to make sure that the expression is entered correctly. Type

```
In[20]:=  Expand[%]

Out[20]=              2         3         4       5
          8 x - 24 x  + 22 x  - 12 x  + 9 x
```

As mentioned earlier, the percent sign % refers to the output of the command last executed by *Mathematica*—in this case, the expression $(3x-2)^2(x^3+2x)$.

To factor the polynomial x^6-1, type

```
In[21]:=  x^6-1
Out[21]=        6
          -1 + x

In[22]:=  Factor[%]
Out[22]=                                   2              2
          (-1 + x) (1 + x) (1 - x + x ) (1 + x + x )
```

Another useful command is `Simplify`. For example, to simplify the expression

$$\frac{x^2 - x}{x^3 - x} - \frac{x^2 - 1}{x^2 + x}$$

enter

```
In[23]:=  (x^2-x)/(x^3-x)-(x^2-1)/(x^2+x)
Out[23]=          2           2
            -1 + x      -x + x
          -(-------) + -------
                  2           3
             x + x      -x + x

In[24]:=  Simplify[%]
Out[24]=         1     1
          -1 + - + -----
               x   1 + x
```

The following command will also simplify the same expression.

```
In[25]:=  Simplify[(x^2-x)/(x^3-x)-(x^2-1)/(x^2+x)]
Out[25]=         1     1
          -1 + - + -----
               x   1 + x
```

With this syntax, however, it is harder to keep track of the parentheses in such a long expression. In addition, this command does not display the original expression, and therefore it cannot be checked for typing errors.

Note: Mathematica has an on-line help facility that you invoke by typing `?`. You can obtain help with a specific command by typing a `?` followed by the name of the command. For example, to get help with the `Factor` command, type `?Factor`. Moreover, if are unsure about the name of a command, but you remember that it contains a certain word, for example "Plot," you can find out all the commands containing `Plot` by typing `?*Plot*`.

1.4 Plots

The `Plot` command is best introduced with an example. To plot the graph of

$$y = \frac{2x^2 - 4}{x + 1}$$

over the interval $-6 \leq x \leq 6$, type

```
In[26]:=  Plot[(2x^2 - 4)/(x + 1), {x, -6, 6}]
```

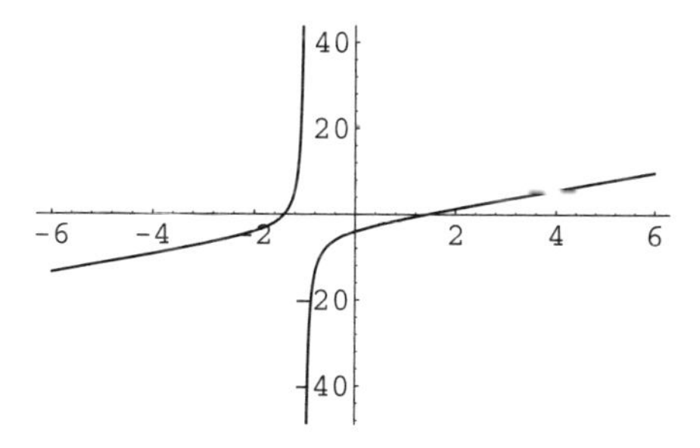

The scale on the y-axis is much different from the scale on the x-axis because of the large function values when x is close to -1 (where the function becomes undefined). Use the option `PlotRange` to see a specific range of y-values. For example, to view the piece of the graph with $-20 \leq y \leq 20$, enter

```
In[27]:=  Plot[(2x^2 - 4)/(x + 1), {x, -6, 6},
             PlotRange -> {-20, 20}]
```

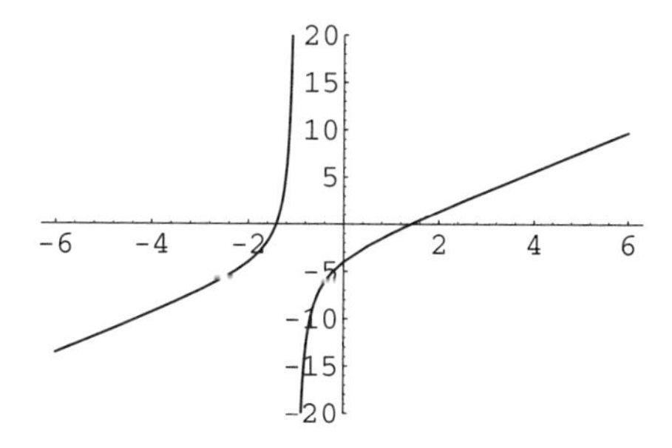

By changing the plot range, you can view different aspects of the graph. For example, changing the x-range to $[-50, 50]$ and the y-range to $[-100, 100]$ displays the graph for larger values of x, where the graph of the function approaches the line $y = 2x - 2$ as a skewed asymptote. However, with such large values of x, the vertical asymptote at $x = -1$ becomes obscured.

```
In[28]:=  Plot[(2x^2 - 4)/(x + 1), {x, -50, 50},
              PlotRange -> {-100, 100}]
```

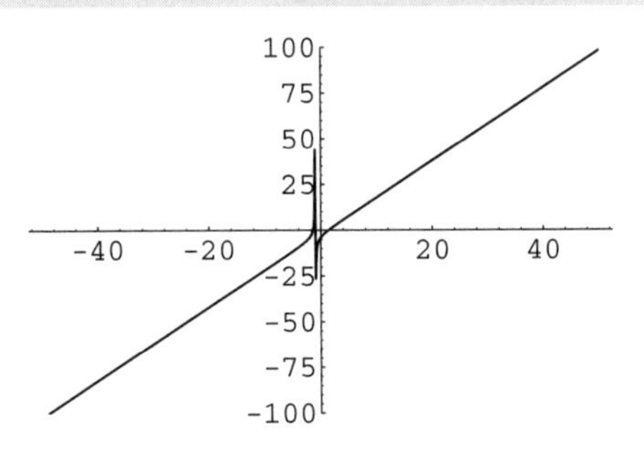

Note that the scale on the y-axis is different from the scale on the x-axis. You can force the scale on the x- and y-axes to be the same by specifying the option setting `AspectRatio -> Automatic`

```
In[29]:=  Plot[(2x^2 - 4)/(x + 1), {x, -6, 6},
              AspectRatio -> Automatic]
```

You can graph more than one expression on the same plot by specifying a list of the expressions as the first argument to `Plot`. The expressions must be enclosed in braces { }. For example, to plot the graphs of x^2 and x^3 on the same coordinate axis over the range $-2 \leq x \leq 3$, type

```
In[30]:=  Plot[{x^2, x^3}, {x, -2, 3}]
```

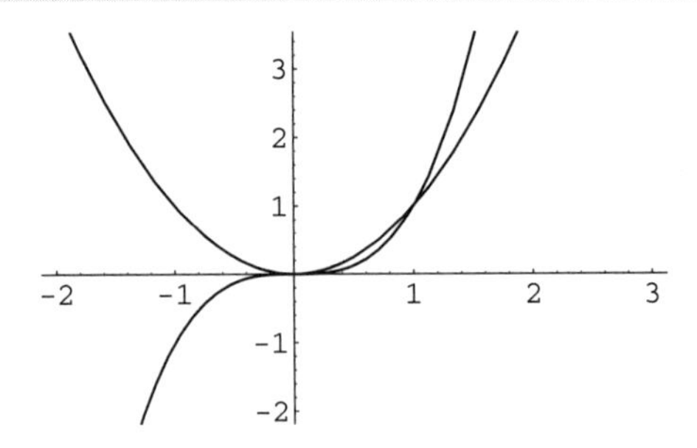

Other options for `Plot` will be discussed in Section 2.4.

1.5 Limits

To compute the limit $\lim_{x \to 2} \dfrac{x^2 - 4}{x - 2}$, enter

```
In[31]:=  Limit[(x^2 - 4)/(x - 2), x -> 2]

Out[31]=  4
```

The `Limit` command computes the limit.

It is instructive to plot the expression, in addition to calculating its limit.

```
In[32]:=  Plot[(x^2 - 4)/(x - 2), {x, 0, 4}]
```

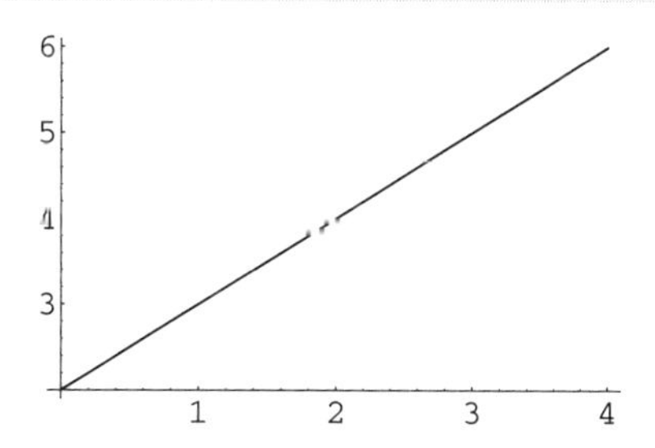

From the graph of $y = (x^2 - 4)/(x - 2)$, it is clear that $\lim_{x \to 2}(x^2 - 4)/(x - 2) = 4$, since y approaches 4 as x approaches 2. The graph of $y = (x^2 - 4)/(x - 2)$ should contain a "hole" at the point $(2, 4)$ because the expression is not defined at $x = 2$. *Mathematica* prints warning messages when it finds an expression that is undefined. Notice that *Mathematica* does not always print warning messages. For example, no warning messages are printed when you try, for example, to plot $g = (x^2 - 3.9601)/(x - 1.99)$ for x in the range $[0, 4]$. The reason is that *Mathematica* plots an expression by computing many points that belong to the graph, and then connects these points by straight-line segments. If one of the points that *Mathematica* uses for a plot happens to be a point at which the expression is undefined, then a warning message will be printed.

For some expressions, the limit at a certain point may depend on the direction from which you approach that point. The right limit is obtained by computing the value of an expression as you decrease toward the limit. You can force `Limit` to compute the right limit by specifying the option setting `Direction -> -1`. Similarly, the left limit is obtained by increasing toward the limit. You can force `Limit` to compute the left limit by specifying the option setting `Direction -> 1`.

```
In[33]:=  f = (x^2 - 4)/(x^2 - 5x + 6)

Out[33]=        2
           -4 + x
          ------------
                    2
          6 - 5 x + x

In[34]:=  Plot[f, {x, 0, 4}, PlotRange -> {-50, 50}]
```

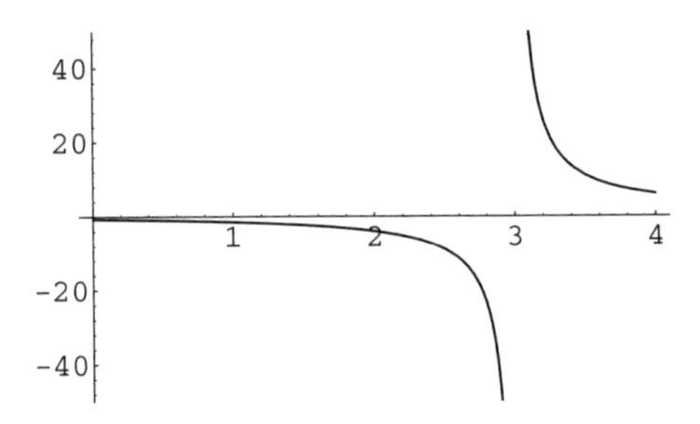


```
In[35]:=  Limit[f, x -> 3, Direction -> -1]

Out[35]=  Infinity

In[36]:=  Limit[f, x -> Infinity]

Out[36]=  1
```

1.6 Summary

- *Mathematica* does everything that a graphing calculator will do.
- Arguments to all *Mathematica* function must be in brackets; for example, $\sin x$ must be written as `Sin[x]`.
- *Mathematica* executes arithmetic commands in a predefined order. If you are not sure of the order in which *Mathematica* will evaluate an expression, use parentheses.
- You can use a space or an asterisk, *, to designate multiplication. Juxtaposition of a number followed by a symbol or parenthesized expressions is a synonym for multiplication: so, for example, `(x+4)(x+2)` is viewed as a product.

- Most calculators store numbers only as floating-point decimals, with a mantissa containing a preselected number of digits of accuracy and an exponent. *Mathematica* has an alternative exact mode: it treats the fraction 1/3 not as a decimal 0.333333 to any number of 3's, but instead as the number 1 divided by the number 3.

- When you calculate a function of an exact number, such as 2 or 1/3, *Mathematica* gives an exact answer. Sometimes, *Mathematica* will parrot back the expression that you entered. When given the input `Sin[2]`, for example, *Mathematica* returns the same expression because that is the exact value of that expression.

- You can calculate floating-point decimal approximations from exact values using the function `N`.

- You need not retype intermediate results in subsequent calculations. *Mathematica* provides two alternatives: the assignment command, `=`, which can be used to assign a name to a value; and percent, `%`, to reference the result just calculated.

- Once a name or symbol is assigned to a number or algebraic expression, any statement that contains that name treats it as a synonym for the number or expression itself.

- Once an assignment has been made, *Mathematica* remembers that assignment until it is told otherwise. Remember that this assignment stays in effect until the name is reassigned or cleared.

- You can check whether a symbol is free or unassigned by using the on-line help, e.g., ?*name*, where *name* is the name of the symbol. You can clear a symbol using the function `Clear` when you no longer want any values to be assigned to that symbol.

- You can use ?*command* to learn about the purpose of a command and how to call it. Use ?**name** to see the names of the commands that contain the characters *name*. The asterisk, `*`, is a wildcard character that matches zero or more characters.

- You can use the commands `Expand`, `Factor`, and `Simplify` to perform algebraic manipulations.

- You can produce a graph of an expression using `Plot`. You can change how a graph appears by specifying options in addition to the required arguments.

- If you are using the Notebook Front End version of *Mathematica*, you can see in the lower edge of a window the approximate coordinates of a point by clicking the mouse button on a graph, then holding down the ⟨COMMAND⟩ key and positioning the pointer above the point on the graph.

- You can compute the limiting value of an expression, call `Limit[`*expr*`, `*var*`->` *value*`]`, e.g., `Limit[Sin[x]/x, x -> 0]`.

1.7 Exercises

1. Assign the variable name *a* to the number $2\pi/5$, and then use `N` to compute decimal approximations for a^2, $1/a$, $\sqrt{a}$, $a^{1.3}$, $\sin(a)$, and $\tan(a)$.

2. To calculate a^2 to 20 significant digits, type `N[a^2, 20]`. Repeat exercise 1 with 20 significant digits.

3. Expand the following expressions:

   (a) $(x^2 + 2x - 1)^3(x^2 - 2)$

   (b) $(x + a)^5$

   Note: If you have done exercise 1 or 2, the label `a` already has a value assigned to it. Recall that you should unassign this value by typing `a =.` or `Clear[a]`.

4. Factor the expression $x^2 + 3x + 2$. What happens if this expression is changed to $x^2 + 3.0x + 2.0$?

5. Try factoring $x^2 + 3x - 11$.

6. Factor $x^8 - 1$.

7. Simplify
$$\frac{2x^2}{x^3 - 1} + \frac{3x}{x^2 - 1}$$

8. Plot the graph of $y = \tan(x)$. Experiment with the x- and y-ranges to obtain a reasonable plot of one period of $y = \tan(x)$.

9. Plot the graph of $y = (3x^2 - 2x + 1)/(x - 1)$ over a small interval containing $x = 1$, for example, $0 \le x \le 2$. Experiment with the y-range to obtain a reasonable plot. What happens to the graph near $x = 1$?

   Now plot the same expression over a large interval, such as $-100 \le x \le 100$. Note that the behavior of the graph near $x = 1$ is no longer apparent. Why do you think this happens?

10. Plot the expressions $\sin(x)$, $\sin(2x)$, and $\sin(4x)$ over the interval $0 \leq x \leq 2\pi$ on the same coordinate axes. Now plot the same expressions over the interval $0 \leq x \leq 4\pi$.

11. Plot the graph of the expression $(x^3+1)/(x+1)$ over the interval $-2 \leq x \leq 0$. From this graph, estimate the value of $\lim_{x \to -1} (x^3+1)/(x+1)$. Now compute this limit with *Mathematica*'s `Limit` command.

12. Plot the graph of the expression $\sin x/x$ over the interval $-1 \leq x \leq 1$. From this graph, estimate the value of $\lim_{x \to 0} \sin x/x$. Now compute this limit with Mathematica's `Limit` command.

13. Compute 7!.

14. Use `N[Pi, 40]` to give the first 40 digits of π.

15. Compute the exact and floating-point values of $\sin(\pi/4)$.

16. Compute the exact and floating-point values of $\sin(1)$.

17. Compute the number of seconds in 1 year, showing the units in your product as each factor is entered.

18. Just as `Factor` will decompose a polynomial into irreducible factors, there is also an integer factor command that gives the prime decomposition of an integer. Use the integer factor command `FactorInteger` to show that $2^{23} - 1$ is not prime.

19. Describe what happens when the command `Expand` is applied to $(a+b)/c$.

20. Use `Expand` to change $\ln(ab/c)$. Practice the technique of entering the expression and checking its *Mathematica* output to see that you entered it correctly. Then, go back and insert the `Expand` command on the same line.

21. Factor the expression $e^{2x} - 1$, by using first `Expand`, then `Factor`.

22. Plot the expressions $-x/2 + 5/2$ and $-3x + 5$ on the same graph. Limit the plot to x-values between 0 and 2. Specify the option setting `AspectRatio -> Automatic` to render a graph in which one unit in the x-direction is drawn the same size as one unit in the y-direction. If you are using the Notebook Front End, find the coordinates of the point of intersection of the lines $y = -x/2+5/2$ and $y = -3x + 5$ as follows.

    (a) Select the graph by holding down the button on your mouse while the pointer on the graph.

(b) Hold down the $\langle$COMMAND$\rangle$ key and move the pointer to the point of intersection of the lines $y = -x/2 + 5/2$ and $y = -3x + 5$. The coordinates of the position of the pointer will be shown in the bottom edge of the window.

2 Solving and Visualizing Equations

This chapter introduces expressions, functions, and equations in *Mathematica*. It also describes the syntax necessary to plot these objects, as well as the syntax involved with solving equations.

2.1 Expressions

In the previous chapter, we saw that we can assign a name to a value (such as `area = 25 Pi`). In a similar way, names can be assigned to expressions that may contain variables. Name assignment is useful when you need to refer to a complicated expression repeatedly. For example, suppose that we want to perform several operations on the expression $2x^3 - 5x^2 + x + 2$. Instead of repeatedly typing this expression, we can assign a name to the expression, such as `f`, and then refer to it in the future by typing `f`.

To assign the name `f` to the expression $2x^3 - 5x^2 + x + 2$, enter

```
In[1]:=  f = 2 x^3 - 5 x^2 + x + 2
Out[1]=              2     3
         2 + x - 5 x  + 2 x
```

The name `f` now is replaced by the *Mathematica expression*. This expression can be factored.

```
In[2]:=  Factor[f]

Out[2]=  (-2 + x) (-1 + x) (1 + 2 x)
```

To plot `f` over the interval $-2 \leq x \leq 3$, type

```
In[3]:=  Plot[f, {x, -2, 3}];
```

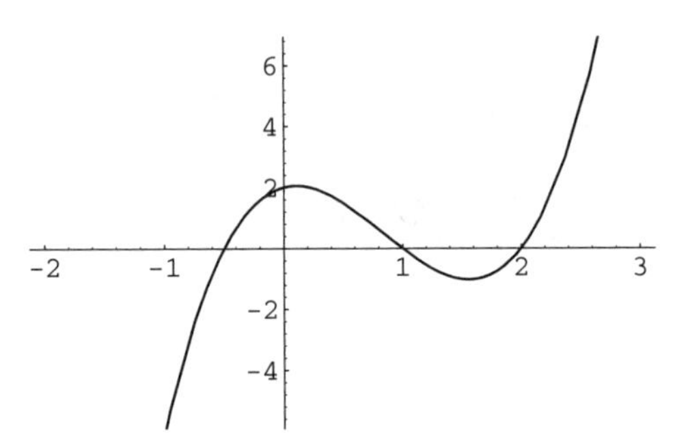

The semicolon at the end of the line `Plot[f, {x, -2, 3}];` suppresses the output line *Out[n]* = `-Graphics-`, which would otherwise appear.

Expressions can involve more than one variable. For example, assign to the variable `volume` the volume of a cylinder of radius `r` and height `h`.

```
In[4]:= volume = Pi h r^2

Out[4]=       2
        h Pi r
```

In this case, the volume involves two variables r and h, representing the radius and height of the cylinder. There is no limit to the number of letters that can be in a name. We recommend using names that indicate their purpose, as is often done in *Mathematica* itself. Part of the reason for the success of *Mathematica* is that people who are not familiar with *Mathematica* can often figure out what a command will do based on that command's name.

There are two ways to substitute values into expressions. The first way is to assign values to the variables. For example, enter

```
In[5]:= x = 6

Out[5]= 6
```

After the assignment is made, the value $x = 6$ is automatically substituted into the expression `f`.

```
In[6]:= f

Out[6]= 260
```

Clear the value assigned to `x` by typing `Clear[x]`, so that the variable or name `x` is free for future use.

The second way to substitute values into variables in an expression is to use `/.`, which is shorthand notation for the command `ReplaceAll`. For example, to substitute the value $x = 6$ into the expression $f = 2x^3 - 5x^2 + x + 2$, enter

```
In[7]:=  f /. x -> 6

Out[7]=  260
```

Think of `/.` as meaning "*given that*" and `->` as meaning "*goes to*." So the expression `f /. x -> 6` can be interpreted as `f` given that `x` goes to `6`.

There is an important difference between the two methods for substituting values into expressions. When you use the first method, the value of `f` is changed to 260 (so, if you try to plot `f`, for example, you will get a horizontal line at $y = 260$). With the second method, you obtain the result 260. However, the values of `f` and `x` are not changed.

Other variables can be substituted for `x`. Substitute $x = a + h$.

```
In[8]:=  f /. x -> a + h

Out[8]=
                               2             3
         2 + a + h - 5 (a + h)  + 2 (a + h)
```

Again, the value of `f` is not changed. To change the value of `f`, enter the following assignment statement: `f = f /. x -> a + h`.

You can enter more than one replacement by specifying a list of the substitutions enclosed in braces, { }. For example, to substitute $r = 2$ and $h = 5$ into the expression for the volume of a cylinder, enter

```
In[9]:=  volume /. {r -> 2, h -> 5}

Out[9]=  20 Pi
```

2.2 Functions

In the preceding discussion, we used the `ReplaceAll` command to substitute values into expressions. For example, to substitute $x = 6$ into the expression $f = 2x^3 - 5x^2 + x + 2$, enter the command `f /. x -> 6`. However, in mathematics, the notation $f(6)$ is more commonly used to indicate the value of the function f at $x = 6$. This syntax will not make any sense in *Mathematica* unless f is first entered as a *function*, instead of an *expression*, and the parentheses are replaced by square brackets: `f[6]`. In this section, we describe how to define you own functions in *Mathematica*.

Before entering this function, remember to clear the definition for `f`, so that the symbol `f` does not get replaced by the value that we previously assigned to it, since a new definition does not necessarily overwrite the previous one. A user-defined function is of the form *function*[*arg*_] := *body*. The symbol *function* is the name that you assign to the function. The argument is enclosed in square brackets. The argument can be a fixed *Mathematica* expression or a symbol followed by a single underscore. The advantage of using a symbol followed by an underscore is that this method generalizes the definition so that the definition can apply to any expression. Underscores appear on the left side of the assignment. The argument names on the left-hand side of the equal sign appear on the right-hand side without the underscore.

To define the function $f(x) = 2x^3 - 5x^2 + x + 2$, type

```
In[10]:=  Clear[f]
          f[x_] := 2 x^3 - 5 x^2 + x + 2
```

With this definition, the name f is used to refer to the argument passed to f. For example, when you call f[6], the variable x is replaced by 6 in the expression $2x^3 - 5x^2 + x + 2$.

```
In[11]:=  f[6]

Out[11]=  260
```

This function can also be evaluated at $x = a + h$.

```
In[12]:=  f[a + h]

Out[12]=
                               2             3
          2 + a + h - 5 (a + h)  + 2 (a + h)
```

To plot the function $f(x) = 2x^3 - 5x^2 + x + 2$ over the interval $-2 \leq x \leq 3$, enter

```
In[13]:=  Plot[f[x], {x, -2, 3}];
```

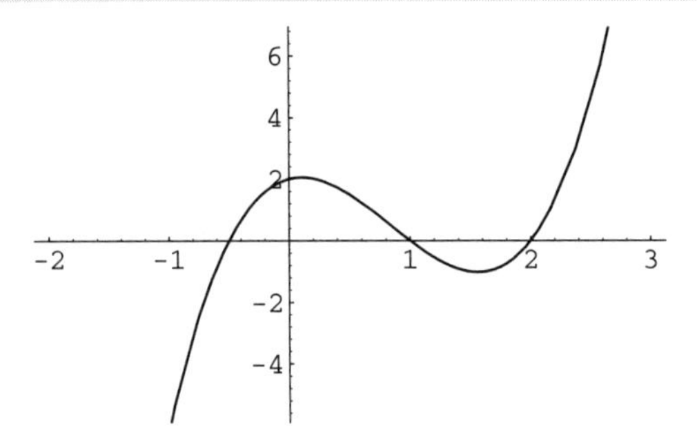

Functions of two or more variables can also be entered in *Mathematica*. For example, to enter into *Mathematica* the function $volume(x, y) = x^2y$ (which represents the volume of a rectangular box with square base of side length x and with height y), type

```
In[14]:=  Clear[volume]
          volume[x_, y_] := x^2  y
```

To evaluate volume at $x = 3$ and $y = 5$, type `volume[3,5]`.

2.3 Solution of Equations

Mathematica has three commands for solving equations: `Solve`, `NSolve`, and `FindRoot`. `Solve` is used for calculating exact or symbolic solutions. `NSolve` and `FindRoot` are used for calculating numerical approximations to the solutions. To solve the quadratic equation

$$x^2 + 2x - 1 = 0$$

enter

```
In[16]:=  Solve[x^2 + 2 x - 1 == 0, x]

Out[16]=          -2 - 2 Sqrt[2]          -2 + 2 Sqrt[2]
          {{x -> --------------}, {x -> --------------}}
                        2                       2
```

The notation == is used for denoting equality and for specifying an equation in *Mathematica*.

You can assign a name to a result. Here, we name the result `solution`.

```
In[17]:=  solution = Solve[x^2 + 2 x - 1 == 0, x]

Out[17]=          -2 - 2 Sqrt[2]          -2 + 2 Sqrt[2]
          {{x -> --------------}, {x -> --------------}}
                        2                       2
```

You can refer to the first and second solutions by typing `solution[[1]]` and `solution[[2]]`. The braces in `solution` designate a list, and, for any list, double square brackets enclosing a positive integer k extracts the kth entry in the list.

```
In[18]:=  solution[[1]]

Out[18]=        -2 - 2 Sqrt[2]
          {x -> --------------}
                      2


In[19]:=  solution[[2]]

Out[19]=        -2 + 2 Sqrt[2]
          {x -> --------------}
                      2
```

If you expect to refer to the expression $x^2 + 2x - 1$ more than once, you might want to assign a name to it.

```
In[20]:=  expr = x^2 + 2 x - 1

Out[20]=             2
          -1 + 2 x + x
```

You can now check the solution by entering `Plot[expr, {x, -3, 3}]` to see that the graph of `f` crosses the x axis at the solutions. Alternatively, you can substitute the values $x = -1 + \sqrt{2}$ and $-1 - \sqrt{2}$ into `expr` (with the commands `expr /. solution[[1]]` and `expr /. solution[[2]]`) to see that the value of `expr` is zero at these solutions.

Names can be given to part of an equation or to an entire equation. After assigning the name `expr` to the expression $x^2 + 2x - 1$, you can solve the equation $x^2 + 2x - 1 = 0$ with the command

```
In[21]:=  solution = Solve[expr == 0, x]

Out[21]=          -2 - 2 Sqrt[2]          -2 + 2 Sqrt[2]
          {{x -> --------------}, {x -> --------------}}
                        2                       2
```

Alternatively, you can assign a name to the entire equation. Here, we assign the name `eq` to the equation.

```
In[22]:=  eq = (x^2 + 2 x - 1 == 0)

Out[22]=             2
          -1 + 2 x + x  == 0
```

It is not necessary to enclose the equation in parentheses. However, doing so may help you to understand how *Mathematica* interprets the input.

Now we can solve this equation:

```
In[23]:=  Solve[eq, x]

Out[23]=          -2 - 2 Sqrt[2]          -2 + 2 Sqrt[2]
          {{x -> --------------}, {x -> --------------}}
                        2                       2
```

Naming either the equation or the left side of the equation is useful should you want to refer to the expression several times, because then you can use the name instead of reentering the entire expression.

You can solve more than one equation by specifying the list of the equations as the first argument to `Solve`. The equations must be enclosed in braces { }. Consider the following two linear equations:

```
In[24]:=  eq1 = (3 x + 2 y == 1);

In[25]:=  eq2 = (x  + 2 y == 3);
```

Here, we solve these equations simultaneously.

```
In[26]:=  solution = Solve[{eq1, eq2}, {x, y}]

Out[26]=  {{x -> -1, y -> 2}}
```

We obtain the solution as a list of rules for x and y. Although this system of equations has been solved, the values of x and y have not been changed. To assign these values to x and y, use the following command:

```
In[27]:=  {x, y} = {x, y} /. solution[[1]]

Out[27]=  {-1, 2}
```

Why do we need to call `solution[[1]]`, instead of simply `solution`? Because *Mathematica* returns a list of lists of the solutions. The command `solution[[1]]` returns a single list with the solution.

Since we anticipate using x and y again, we clear the values from x and y.

```
In[28]:=  Clear[x, y]
```

Exact solutions to complicated equations can be difficult or impossible to find. For example, enter the following equation:

```
In[29]:=  eq = (x^3 - 2 x^2 + x - 3 == 0)

Out[29]=               2    3
          -3 + x - 2 x  + x  == 0
```

Solving this equation with the command `Solve[eq, x]` will yield three solutions that are complicated (try solving it). To get decimal approximations to these solutions, change the coefficients of the equation from exact integers to floating-point decimals. When `Solve` encounters input that has numbers with decimal points, it assumes that your input is approximate and consequently calculates an approximation to the result.

```
In[30]:=  eq = (x^3 - 2.0 x^2 + x - 3.0 == 0)

Out[30]=                 2    3
          -3. + x - 2. x  + x  == 0
```

Now solving the equation will yield simpler (but only approximate) decimal answers.

```
In[31]:=  Solve[eq, x]

Out[31]=  {{x -> -0.0872797 - 1.17131 I},
            {x -> -0.0872797 + 1.17131 I}, {x -> 2.17456}}
```

The capital I in the result represents $\sqrt{-1}$.

Some equations are impossible to solve exactly. In fact, there is a theorem in mathematics that states that there is no formula (analogous to the quadratic formula) for finding roots of polynomials of degree five or higher. Consider the following equation:

```
In[32]:=  eq = (x^7 + 3 x^4 + 2 x - 1 == 0)

Out[32]=                  4     7
          -1 + 2 x + 3 x   + x   == 0
```

Try to find the exact solutions using *Mathematica*.

```
In[33]:=  Solve[eq, x]

Out[33]=                                 4     7
          {ToRules[Roots[2 x + 3 x   + x   == 1, x]]}
```

This response indicates that *Mathematica* does not know how to solve this equation exactly. In this situation, use *Mathematica*'s NSolve command to calculate approximate solutions.

```
In[34]:=  NSolve[eq, x]

Out[34]=  {{x -> -1.1926 - 0.179308 I},
            {x -> -1.1926 + 0.179308 I},
            {x -> 0.245205 - 0.898052 I},
            {x -> 0.245205 + 0.898052 I}, {x -> 0.441418},
            {x -> 0.726682 - 1.12661 I},
            {x -> 0.726682 + 1.12661 I}}
```

Now consider the equation

```
In[35]:=  eq = (x^2 + 1/x - 1/x^2 == 0)

Out[35]=    -2   1     2
          -x   + - + x   == 0
                 x
```

`NSolve` is intended to be used for finding numerical approximations to the roots of an equation or set of equations. Solving this equation with `NSolve` will yield four solutions: two that are real valued and two that are complex valued.

```
In[36]:= NSolve[eq, x]

Out[36]= {{x -> -1.22074}, {x -> 0.248126 - 1.03398 I},
           {x -> 0.248126 + 1.03398 I}, {x -> 0.724492}}
```

With `Plot`, you can see the two real-valued solutions.

```
In[37]:= Plot[x^2 + 1/x - 1/x^2, {x, -2, 2},
              PlotRange -> {-20, 20}]
```

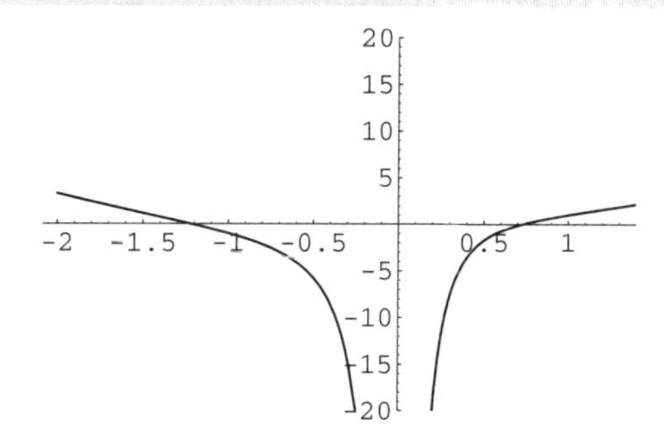

You can also obtain these solutions by calling `FindRoot` twice. You must provide `FindRoot` two arguments: the expression for which you want to find a root, and a value to use when starting the search for a root. Here, we give the starting value $x = 1$.

```
In[38]:= FindRoot[eq, {x, 1}]

Out[38]= {x -> 0.724492}
```

The `NSolve` command finds all the roots of a polynomial. The `FindRoot` command solves more complicated equations.

Example 2.1 Find the equation of the parabola that passes through the points $(-1, 2)$, $(1, -1.5)$, and $(4, 7)$.
Solution. First, enter the general equation of a parabola (defined as a function of x).

```
In[39]:= p[x_] := a x^2 + b x + c
```

The unknown coefficients `a`, `b`, and `c` must be found such that the parabola passes through the points $(-1, 2)$, $(1, -1.5)$, and $(4, 7)$. For the parabola to pass through the point $(-1, 2)$, the equation $p(-1) = 2$ must be satisfied. We assign the name eq1 to this equation.

```
In[40]:=  eq1 = (p[-1] == 2)

Out[40]=  a - b + c == 2
```

The other two equations corresponding to the other two points are entered similarly.

```
In[41]:=  eq2 = (p[1] == -1.5)

Out[41]=  a + b + c == -1.5

In[42]:=  eq3 = (p[4] == 7)

Out[42]=  16 a + 4 b + c == 7
```

We solve these equations and assign a name to the result.

```
In[43]:=  solution = Solve[{eq1, eq2, eq3}, {a, b, c}]

Out[43]=  {{a -> 0.916667, b -> -1.75, c -> -0.666667}}
```

We can substitute these values of a, b, and c into p[x] using /. (an alias for the ReplaceAll command).

```
In[44]:=  expr = p[x] /. solution

Out[44]=                                    2
          -0.666667 - 1.75 x + 0.916667 x
```

Since we defined expr, we can use it to check whether the parabola passes through the desired points by invoking the command Plot[expr, {x, -2, 5}].

The Plot command plots an expression or several expressions. The ImplicitPlot command graphs the solution implied by an equation or set of equations. To use this command, first load the package ImplicitPlot.m that is in the directory (or folder) Graphics with the command

```
In[45]:=  Needs["Graphics`ImplicitPlot`"]
```

Here, we enter the equation $x^2 + y^2 = 1$, and name it eq.

```
In[46]:=  eq = (x^2 + y^2 == 1)

Out[46]=   2    2
          x  + y  == 1
```

Now we use the command ImplicitPlot to graph the equation.

```
In[47]:=  Needs["Graphics`ImplicitPlot`"]
          ImplicitPlot[eq, {x, -1, 1}, {y, -1, 1}];
```

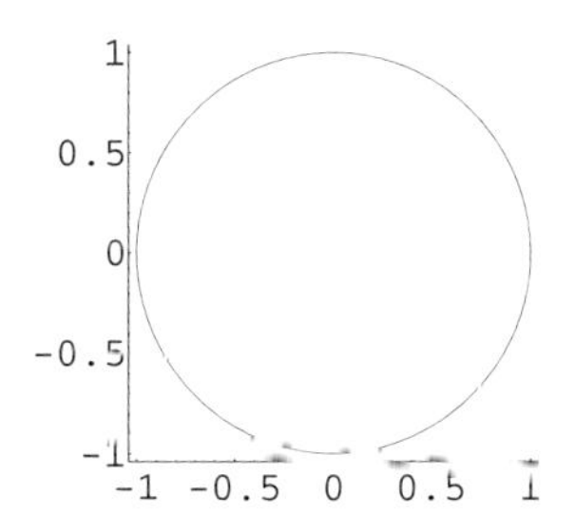

2.4 Visualizing Expressions and Equations

We have shown you how to obtain the solutions of equations. Now we will show how you can graph solutions of implicit equations and expressions. Fortunately, *Mathematica* has a repertoire of built-in functions for producing graphs. Here is a table of useful commands. Later in this section, we show how to use some of these functions.

Plot[*expr*, {x, $x_{\min}$, $x_{\max}$}]
: Plots *expr* over the interval $x_{\min} \leq x \leq x_{\max}$.

Plot[*expr*, {x, $x_{\min}$, $x_{\max}$}, PlotLabel -> "*plot title*"]
: Specifies the label *plot title* to a plot of *expr* over the interval $x_{\min} \leq x \leq x_{\max}$.

Plot[*expr*, {x, $x_{\min}$, $x_{\max}$}, PlotPoints -> *n*]
: Specifies that at least n sample points are to be taken and used in graphing *expr* over the interval $x_{\min} \leq x \leq x_{\max}$. The option PlotPoints changes the resolution (the number of data points) in a plot.

Plot[*expr*, {x, $x_{\min}$, $x_{\max}$}, PlotRange -> {$y_{\min}$, $y_{\max}$}]
: Plots *expr* over the interval $x_{\min} \leq x \leq x_{\max}$ restricting the displayed values of y in the range $y_{\min} \leq y \leq y_{\max}$.

Plot[{$expr_1$, $expr_2$}, {x, $x_{\min}$, $x_{\max}$}]
: Plots the graphs of $expr_1$ and $expr_2$ over the interval $x_{\min} \leq x \leq x_{\max}$ on the same coordinate axis.

ImplicitPlot[*eq*, {x, a, b}, {y, c, d}]
: Draws a graph of a set of points that satisfy the equation *eqn*.

`ImplicitPlot[`*eq*`, {`*x*`, `*a*`, `*b*`}, {`*y*`, `*c*`, `*d*`}, PlotPoints -> `*n*`]`
: Draws a graph of a set of points that satisfy the equation *eqn* using at least n points. If you do not specify the option, `PlotPoints` assumes its default value of 25 points.

`ParametricPlot[{`*fx*`, `*fy*`}, {`*u*`, `*a*`, `*b*`}]`
: Gives a parametric plot of a curve parametrized by the variable *u*.

`Plot3D[`*expr*`, {`*x*`, `*a*`, `*b*`}, {`*y*`, `*c*`, `*d*`}]`
: Gives a three-dimensional plot of the expression *expr*.

`ParametricPlot3D[{`*fx*`, `*fy*`, `*fz*`}, {`*u*`, `*a*`, `*b*`}]`
: Gives a three-dimensional space curve parametrized by a variable *u*.

`ParametricPlot3D[{`*fx*`, `*fy*`, `*fz*`}, {`*u*`, `*a*`, `*b*`}, {`*v*`, `*c*`, `*d*`}]`
: Gives a three-dimensional space surface parametrized by the variables *u* and *v*.

`ListPlot[{`y_1`, `y_2`, ..., `y_n`}]`
: Plots a list the n values $y_1, y_2, \ldots, y_n$ where the x coordinates for these points are taken to be $1, 2, \ldots n$.

`ListPlot[{{`x_1`, `y_1`}, {`x_2`, `y_2`}, ..., {`x_n`, `y_n`}}]`
: Plots a list of the n values $[x_1, y_1], [x_2, y_2], \ldots [x_n, y_n]$.

`ListPlot[`*data*`, PlotJoined -> True]`
: Plots a list of values with connecting line segments. The data may be a list of y values or pairs of x and y values.

`Show[`*g1*`, `*g2*`]`
: Displays two graphs *g1* and *g2*.

Adaptive Sampling

Mathematica graphs a function or an expression by sampling the expression at points along the plotting interval and connecting the sampled points with straight-line segments. In the graph on the left, the vertical lines show where *Mathematica* takes samples when plotting the expression $(x-1)(x-3)(x-4)$ over the interval $[0.5, 4.5]$. The graph on the right shows the curve as *Mathematica* draws it.

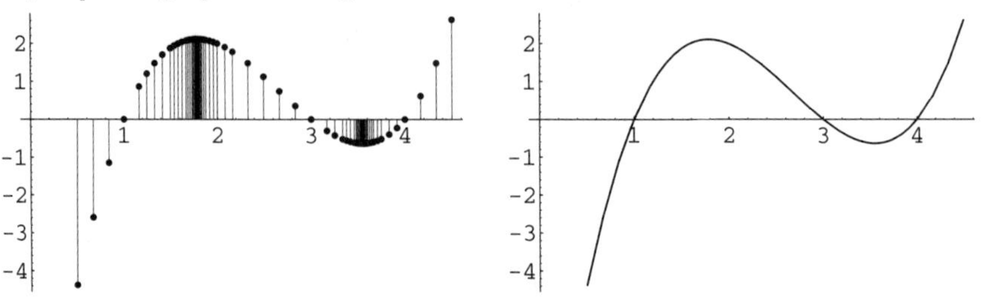

Computing the value of an expression at many points can be time consuming. To avoid computing more points than necessary, *Mathematica* uses a sampling algorithm that adapts itself to the shape of the graph that *Mathematica* is drawing. The algorithm starts sampling on a set of x values equally spaced along the interval over which the function is being plotted. The option `PlotPoints` specifies the number of initial samples. For a two-dimensional plot, the default number of points initially sampled is 25. In intervals where the graph is essentially linear, the initial sample points are sufficient, but in curved sections of the graph, *Mathematica* computes additional sample points. In this way, the algorithm spends extra computing time on the parts of the graph that require it, but does not waste time computing points that are not needed.

Although the adaptive-sampling algorithm can produce reasonable plots with a minimum number of sample points, it can also produce poor renditions. If *Mathematica* samples a plot too infrequently, the plot may show the wrong function—an "aliased" version of the desired one.

Example 2.2 Plot the graph of the expression $f = x + \sin(2\pi x)$ over the interval $0 \leq x \leq 2\pi$. Here is a list of the 25 initial sample points computed by *Mathematica*:

```
In[49]:=  Table[{x, x + Sin[2 Pi x]}, {x, 0, 24}]

Out[49]=  {{0, 0}, {1, 1}, {2, 2}, {3, 3}, {4, 4}, {5, 5},
            {6, 6}, {7, 7}, {8, 8}, {9, 9}, {10, 10},
            {11, 11}, {12, 12}, {13, 13}, {14, 14}, {15, 15},
            {16, 16}, {17, 17}, {18, 18}, {19, 19}, {20, 20},
            {21, 21}, {22, 22}, {23, 23}, {24, 24}}
```

Note that these points lie along the line $y = x$. When you enter the following command, *Mathematica* draws a straight line:

```
In[50]:=  Plot[x + Sin[2 Pi x], {x, 0, 24}];
```

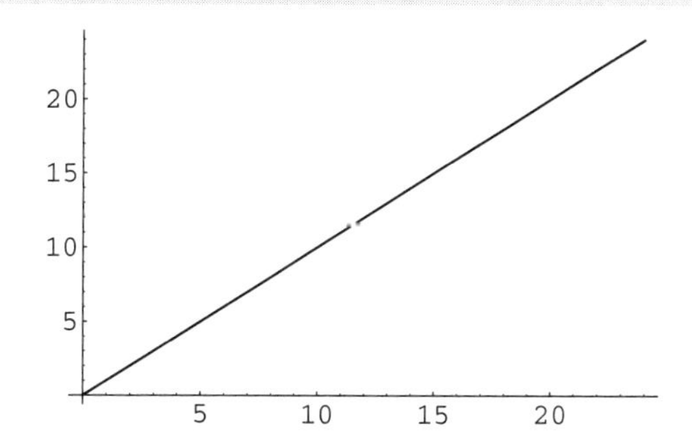

Sampling either at a lower or at a higher frequency, you can obtain a more faithful rendition of the function.

```
In[51]:=  Plot[x + Sin[2 Pi x], {x, 0, 24}, PlotPoints->50];
```

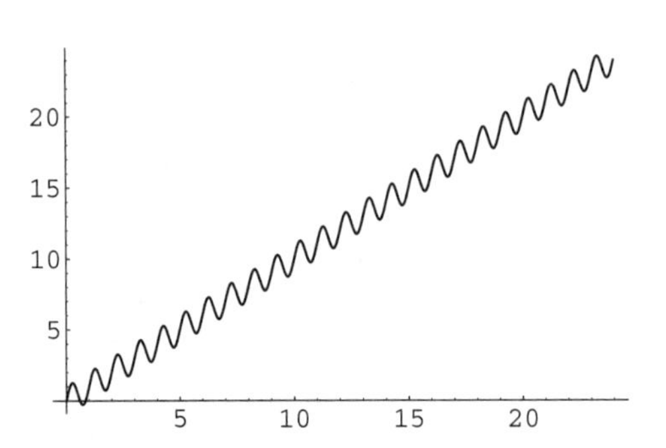

Be aware that the larger the value of PlotPoints, the longer it will take *Mathematica* to compute sampling points and to draw the graph.

In a similar manner, you can increase the resolution of the ImplicitPlot command by adding the option PlotPoints -> n, where n specifies the number of grid points along the x- and y-axes, or PlotPoints -> $\{m,\ n\}$, where m specifies the number of grid points along the x-axis and n specifies the number of grid points along the y-axis. The default value of both m and n is 25. Keep the values of m and n from getting too large so that the computer does not get overloaded. A safe limit for m and n is 50.

Some Examples

Here are some simple examples. Let's start by graphing $e^{-x}\cos x$.

```
In[52]:=  Plot[Exp[-x] Cos[5x], {x, 0, 3}];
```

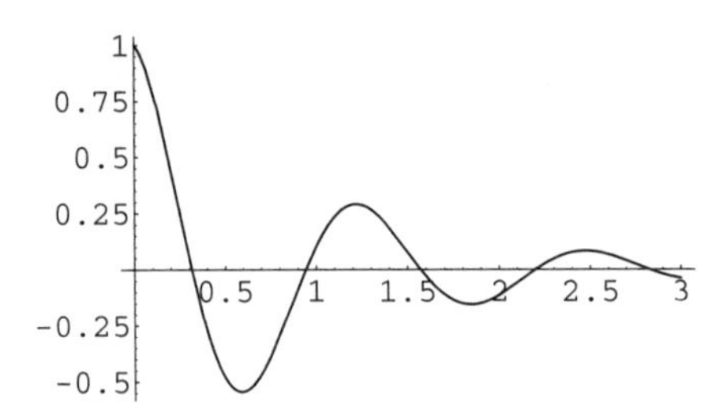

Three-Dimensional Plots

We have discussed expressions and functions of two or more variables. You can obtain a plot of a function or an expression of two variables by using the Plot3D command.

Example 2.3 Plot the expression $\sin(2x + y)$ over the region $-5 \leq x \leq 5, -5 \leq y \leq 5$.

Solution. Now let's use the command Plot3D.

```
In[53]:= expr = Sin[2 x + y];
         Plot3D[expr, {x, -5, 5}, {y, -5, 5}];
```

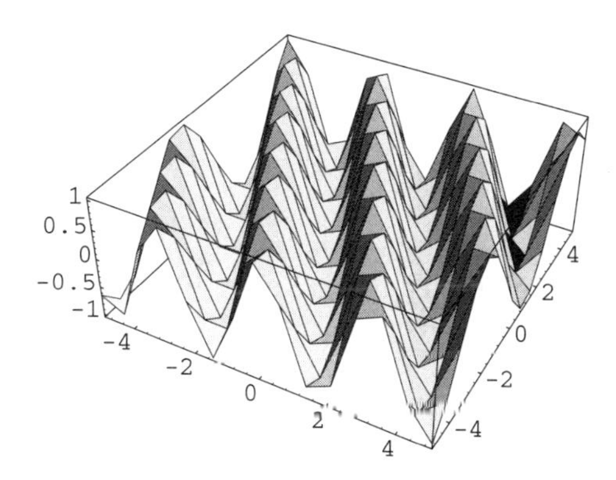

As with Plot and ImplicitPlot, you can increase the resolution of this graph by adding the option PlotPoints -> 25 (the default is 15).

You can view the graph from a different perspective by changing the option setting for ViewPoint. The ViewPoint can be thought of as the point from which you view the object. The ViewPoint is specified relative to a bounding box whose sides are of length 1 and whose center is at $\{0, 0, 0\}$. If you want to view the graph from above, set the ViewPoint to {0,0,2}. You can also select a ViewPoint using the *3D ViewPoint Selector*, which can be accessed from the *Prepare Input* submenu in the *Action* menu. Using the *3D ViewPoint Selector*, you can select a value for ViewPoint interactively by positioning a cube with your mouse.

Here, we define f as a function of two variables.

```
In[55]:= Clear[f]
         f[x_, y_] := Sin[2 x + y]
         Plot3D[f[x, y], {x, -5, 5}, {y, -5, 5}];
```

Plotting of a List

Often, problems involve data rather than functions or expressions. For example, to plot the data set $(1, 2), (-2, 3), (4, 4)$, use the command ListPlot, which is intended for plotting lists of data.

```
In[58]:= ListPlot[{{1,2}, {-2,3}, {4,4}}];
```

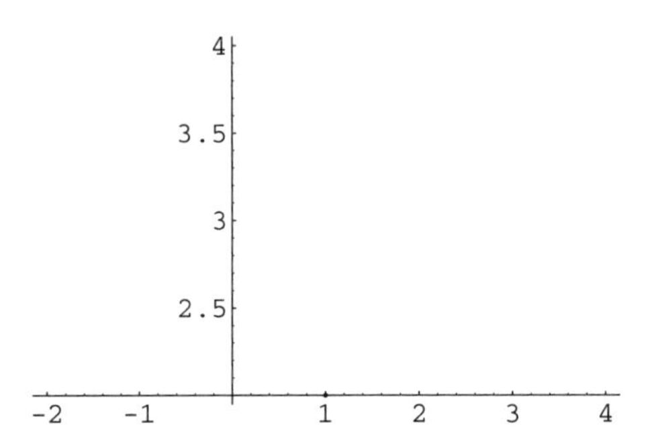

Specify the option setting `PlotJoined -> True` to draw this data set with connecting line segments.

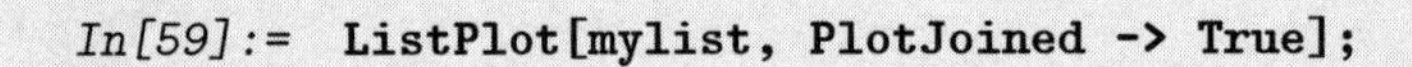

```
In[59]:=  ListPlot[mylist, PlotJoined -> True];
```

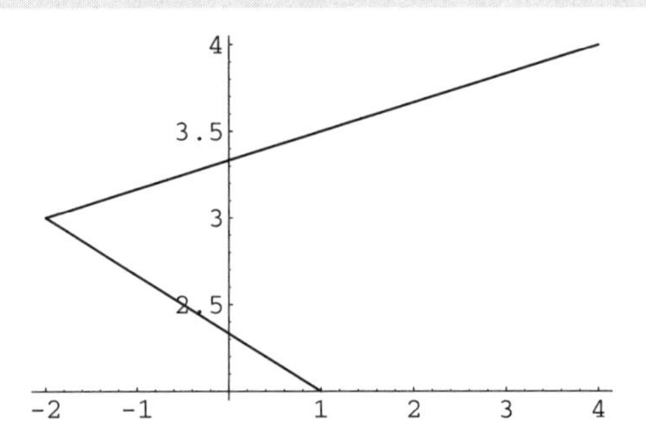

Display of Several Plots

As mentioned in Chapter 1, you can obtain a plot of two or more expressions or functions over the same interval by enclosing the expressions with braces { }. Use the `Show` command to obtain a plot of two or more expressions over *different* intervals.

Example 2.4 Plot the expression $f = x^2$ over the interval $-1 \leq x \leq 1$, and plot the expression $g = 2x - 3$ over the interval $0 \leq x \leq 2$.
Solution. First define these expressions in *Mathematica*. Then, assign their individual plots to variable names—say, `p1` and `p2`.

```
In[60]:=  f[x_] := x^2
          g[x_] := 2 x - 3
          p1 = Plot[f[x], {x, -1, 1}];
          p2 = Plot[g[x], {x, 0, 2}];
```

Note that the graphs labeled `p1` and `p2` are over different ranges. We combine the two graphs with the command `Show`.

```
In[64]:= Show[p1, p2];
```

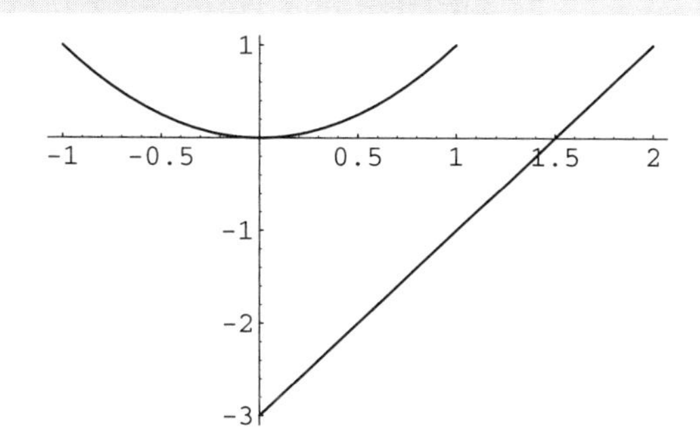

2.5 Summary

- You can define functions using `f[x_] := body`. You can define equations using ==, as in `x + y == 0`.
- You can replace a value in an expression using `/.` and a rule—for example, `x + y /. x -> 3`.
- The assignment command globally changes values; `/.` (an alias for the `ReplaceAll` command) affects only screen output from a given expression. You can combine the two to save the screen output for later use.
- You can plot a *Mathematica* expression by using the `Plot` command.
- You can use the `ImplicitPlot` command to graph equations. Load package `ImplicitPlot.m` by calling `Needs["Graphics`ImplicitPlot`"]` before calling `InplicitPlot`.
- `N` converts an exact number to floating-point decimal format; `Solve`, `NSolve`, and `FindRoot` solve equations.
- `Solve` attempts to produce exact answers if the input equation contains exact values; `NSolve` returns approximations to the roots.
- You can use `Solve` on linear equations, including systems of equations, and quadratics with exact coefficients equations where the answer will contain variables. You can use `NSolve` to obtain numerical approximations to the solutions. Be aware that `Solve` is able to solve some, but not all, problems.
- Except in the case of polynomials, `NSolve` may not find all solutions. In such cases, you can use `FindRoot`. If you are using the Notebook Front End version of *Mathematica*, you can see in the lower edge of a window the approximate

coordinates of the solutions by clicking the mouse button on a graph, then holding down the ⟨COMMAND⟩ key and positioning the pointer above the point on the graph.

- You can use the command `Show` to display two graphs on the same coordinate axes.

- You can use `Plot` command options to adjust the display of plots. To obtain a list of the available options with their default values, enter `Options[Plot]`.

2.6 Exercises

1. Enter an expression that describes the area of a square in terms of that square's diagonal. Use the variable name `area` for the area of the square and let `d` be the diagonal length. Compute the area of the square when $d = 2$ and $d = 3.7$.

2. Enter an expression that describes the area of a circle in terms of that circle's circumference. Use the variable name `area` for the area of the circle, and let `p` be the circumference. Compute the area of the circle when $p = 2$ and $p = 3.7$.

3. Enter an expression that describes the volume of a cube in terms of that cube's diagonal length. Use the variable name `volume` for the volume of the cube, and let `d` be the diagonal length. (If you have done exercise 1, be careful to clear the variable `d` before you enter this expression.) Compute the volume of the cube when $d = 2$ and $d = 3.7$.

4. Repeat exercises 1 through 3 with the expressions entered as *Mathematica* functions. Evaluate these functions when the independent variable (either d or p) is 2 and when it is 3.7.

5. A large tank initially contains 100 gallons of salt water with a concentration of 1/10 pound of salt per gallon. Then, salt water with a concentration of 1/2 pound of salt per gallon begins to flow into the tank at the rate of 3 gallons per minute.

   (a) Express the total amount of salt t minutes after pouring the salt into the water as an expression in t; assign the name `s` to this amount. Evaluate `s` at $t = 4, 10$, and 50 minutes.

   (b) Express the concentration as an expression of t, and evaluate the concentration at $t = 4, 10$, and 50 minutes.

6. Enter the function that describes the volume of a cylinder of radius r and height h as a function of the two variables r and h.

(a) Evaluate this function at $r = 2$ and $h = 3$.

(b) Compute an expression for the volume of a prism with a height h and with base given by an equilateral triangle of side length s. Enter this expression as a function of the two variables s and h. Compute the volume of the prism when $s = 2.3, h = 4.5$, and when $s = 1.7, h = 7.2$.

7. Enter $f = (x^5 + 1)/(x^2 - 1)$ as an expression. Plot the expression f over the interval $-2 \leq x \leq 0$. Evaluate the limit $\lim_{x \to -1} f$.

8. Enter $f(x) = (x^4 - 16)/(x^2 - 4)$ as a function. Plot the graph of $f(x)$ over the interval $1 \leq x \leq 3$. Evaluate the limit $\lim_{x \to 2} f(x)$.

9. Consider the expression entered in *Mathematica* as `f = x^2-1` Which of the following statements produce meaningful results?

   (a) `Plot[f, {x, -2, 2}]`

   (b) `Plot[f[x], {x, -2, 2}]`

   (c) `Factor[f]`

   (d) `f /. x -> 2`

   (e) `f[x] /. x -> 3`

   (f) `f[2]`

   (g) `Solve[f==0, x]`

   (h) `Solve[f[x] == 0, x]`

10. Repeat exercise 9, but enter f as a *function*, `f[x_] := x^2-1`.

11. Define the functions $f(x) = (x + 2)/(2x + 1)$ and $g(x) = x/(x - 2)$ in *Mathematica*. Find the composition of the two functions by $f(g(x))$. Note that the output is an expression, since each function has been evaluated.

12. Use the `Plot` command to draw the graph of

$$f(x) = \frac{\sqrt{4 - x} + \sqrt{3 + x}}{x^2 - 2}$$

showing only $-5 \leq x \leq 5$ and $-10 \leq y \leq 10$.Note the vertical line segments that appear in the graph. Verify that the asymptotes shown in the graph are not really there. Plot the function f, showing only the range of x and y indicated previously. Find $f(-1), f(2), f(5)$.

13. Form the expression $f = \sqrt{(1/3)x^2 \sin(x + \pi/6)}$. Plot this expression. Click on the endpoints of the pieces to determine the domain of the expression. Use `FindRoot` on $\sin(x + \pi/6) = 0$, with the information that you obtained from

the plot to limit the range, and determine to six decimal places the endpoints of the first connected segment of the domain that lies to the right of $x = 0$.

14. Solve the following equations using the `Solve` command. Check your answers with a plot. In each case, substitute the roots into the expression on the right side of the equation to verify that the roots satisfy the equation.

    (a) $x^2 + 3x + 1 = 0$

    (b) $x^2 + 3.0x + 1.0 = 0$

    (c) $x^3 + x + 1 = 0$

    (d) $x^3 + x + 1.0 = 0$

    (e) $x^2 + 2x + 2 = 0$

    (f) $x^2 + 2.0x + 2.0 = 0$

15. Use the `Plot` and `FindRoot` commands to find all solutions to the following equations over the given interval.

    (a) $\sin^2 x = \cos(3x), 0 \le x \le \pi$

    (b) $\cos^2 x = \sin(3x), 0 \le x \le \pi$

    (c) $8 \cos x = x, -\infty < x < \infty$

16. Solve the system of equations $3x + y = 2$ and $2x - 3y = 7$ for x and y.

17. Find the decimal approximations for all roots of the equation $\dfrac{1}{x^4} - 3 + x^2 = 0$.

18. Find the equation of the cubic that passes through the points $(1, 2.4)$, $(3, 5.6)$, $(4, -2.7)$, and $(7, 4.7)$. Graph your answer to ensure that it passes through these points.

19. Define the expression `expr = x Sin[1/x]` in *Mathematica*. Enter the following plot commands, and examine the behavior of each plot near the origin.

    (a) `Plot[expr, {x, -1, 1}];`

    (b) `Plot[expr, {x, -1, 1}, PlotPoints -> 100];`

    (c) `Plot[expr, {x, -1, 1}, PlotPoints -> 300];`

20. Use the `Plot3D` command to view the graphs of the following expressions:

    (a) $x^2 - y^2$

    (b) $x^2 + y^2$

(c) $x^4 - y^4$

(d) $(x^2 - y^2)(x + y)$

(e) $\cos(x + y)$

21. For each of the following, plot the pair of expressions over the given intervals using `Show` (see Section 2.4).

    (a) $f = x^4 - 2x^2, -2 \leq x \leq 2$, and $g = x^3 - x, -1 \leq x \leq 3$.

    (b) $f = \sin x + \cos x, -2\pi \leq x \leq 0$, and $g = \tan(x)/(\tan(x) + 1), 0 \leq x \leq \pi/2$.

22. The general equation of a circle is

$$x^2 + y^2 + ax + by = c$$

Find the equation of the circle (i.e., find a, b, and c) that passes through the three points $(1, 1)$, $(3, -1)$, and $(6, 4)$.

Hint: It is not possible to graph the circle with a function of the form $y = f(x)$ (why is it impossible?) so it is easier to work with expressions using `/.` to obtain the necessary three equations. Start by assigning the preceding equation to the variable name `eq`. Since the circle is to contain the point $(1, 1)$, the first equation is obtained from the command `eq1 = (eq /. {x -> 1, y -> 1})`. The other two equations are obtained in an analogous fashion using the other two points. Then, use the `Solve` command to solve these three equations for a, b, and c. After substituting these values for a, b, and c into the equation of the circle, plot your answer with the `ImplicitPlot` command. Make sure that you have loaded the package `ImplicitPlot.m` (by invoking `Needs["Graphics`ImplicitPlot`"]`) before you call `ImplicitPlot[eq, {x, -10, 10}, {y, -10, 10}]`.

23. The point of this exercise is that plots can be deceiving. Enter the expression $f = x^3 - x^2 - x + 1.001$. Plot f over the interval $-2 \leq x \leq 2$. From this plot, guess how many real solutions there are to the equation $f = 0$. Now solve the equation $f = 0$ with the `Solve` command. How many solutions did `Solve` return? Now replot f over the interval $0.9 \leq x \leq 1.1$. Does the graph of f cross the x-axis near the point $x = 1$?

24. The point of this exercise is to show how piecewise-defined functions can be entered into *Mathematica*. Consider the function

$$f(x) = \begin{cases} x^2 & \text{if } x \leq 1 \\ 2x + 1 & \text{if } x > 1 \end{cases}$$

This function can be entered into *Mathematica* with the following definitions.

```
Clear[f]
f[x_ /; x <= 1] := x^2
f[x_ /; x >  1] := 2x + 1
```

The notation `/;` is an alias or shorthand notation for the command `Condition`. The rule `f[x_ /; x <= 1] := x^2` specifies that `f[x]` is to be replaced by `x^2` when the condition following the slash and semicolon (`/;`) is met. The function `f` can now be evaluated like any other function. For example, you can obtain the value of `f` at $x = 0.5$ and $x = 2$ by entering `f[0.5]` and `f[2]`. To plot `f` over the interval $-1 \leq x \leq 3$, enter

```
Plot[f[x], {x, -1, 3}]
```

Functions with more than two pieces can be defined in *Mathematica* as well. For example, the function

$$f(x) = \begin{cases} x^2 & \text{if } x \leq 0 \\ 2x + 1 & \text{if } 0 < x \leq 1 \\ -2x & \text{if } x > 1 \end{cases}$$

can be entered as

```
Clear[f]
f[x_ /; x <= 0] := x^2
f[x_ /; 0 <= x <= 1] := 2x + 1
f[x_ /; x >  0] := -2x
```

Try entering and plotting the following functions:

$$f(x) = \begin{cases} \sin(x) & \text{if } x < 0 \\ \cos(x) & \text{if } x \geq 0 \end{cases}$$

$$g(x) = \begin{cases} 2x & \text{if } x \leq 0 \\ \sqrt{x} & \text{if } 0 < x \leq 2 \\ x^2 & \text{if } x > 2 \end{cases}$$

25. To plot a map of Texas, enter the following two lists (named `north` and `south` for the northern boundary and southern boundary of Texas).

```
north = {{0,0},{3,0},{3,4.5},{6,4.5},{6,2.2},{7,2.1},
        {8,1.8},{9,1.9},{10,1.8},{11,1.7},{11,-2.2}};
south = {{0,0},{1,-1.1},{2,-2.5},{3,-2.9},{4,-2.3},
        {5,-2.8},{6,-4.4},{7,-5.8},{8,-6.1},{9,-3.3},
        {10,-2.8},{11,-2.2}};
```

Here, the origin is the western corner of Texas (near El Paso), and the x-axis is the extension of the east-west border between New Mexico and Texas. Each unit represents approximately 69 miles. After entering these lists, load the package `Graphics`MultipleListPlot`` using the `Needs` command, and then type `MultipleListPlot[north, south]`

26. Use *Mathematica* to find the equation of the circle with center at $x = 3$ and $y = 3$ with radius 2. Then, graph the circle in two ways:

   (a) `Solve` the equation for y and assign a name to the solution so that you can refer to it easily. Plot the two solutions on the same graph. (Note that the circle looks more like an ellipse than a circle. How can you remove the distortion?)

   (b) Load the package `ImplicitPlot.m` that is in the directory (or folder) `Graphics` by typing `Needs["Graphics`ImplicitPlot`"]`. Then, make an `ImplicitPlot` of the equation.

3 Differentiation

We start this chapter by computing derivatives by using the `Limit` command. Then, we introduce the *Mathematica* derivative command `D`.

3.1 The Limit of the Difference Quotient

The definition of the derivative of the function $f(x)$ at the point $x = a$ is given by

$$f'(a) = \lim_{h \to 0} \frac{f(a+h) - f(a)}{h}$$

The motivation of this definition is that the derivative of f at $x = a$ should be the slope of the tangent line to the graph of f at $x = a$. The key idea is that the quantity

$$\frac{f(a+h) - f(a)}{h}$$

is the slope of the line segment or chord between the point $(a, f(a))$ and $(a + h, f(a+h))$.

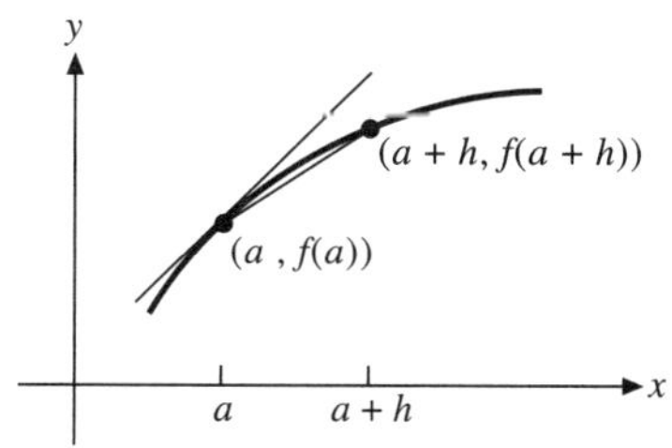

The preceding definition of the derivative reflects the fact that the slope of the tangent line at $x = a$ is the limit of the slopes of these chords as the number h tends to zero.

As an example, consider the function $f(x) = x^3 - 8x$. Define this function in *Mathematica* as shown below and calculate $f'(2)$ by taking the above limit with $a = 2$.

```
In[1]:=  Clear[f]
         f[x_] := x^3 - 8x
         Limit[(f[2 + h] - f[2])/h, h -> 0]

Out[3]=  4
```

The value of $f'(2)$, the slope of the tangent line to f at $x = 2$, is 4. The tangent line also passes through the point $(2, f(2)) = (2, -8)$. Accordingly, its formula is

```
In[4]:=  y = 4 (x - 2) - 8

Out[4]=  -8 + 4 (-2 + x)
```

Now we plot the tangent line and the function on the same coordinate axis.

```
In[5]:=  Plot[{y, f[x]}, {x, 0, 4}];
```

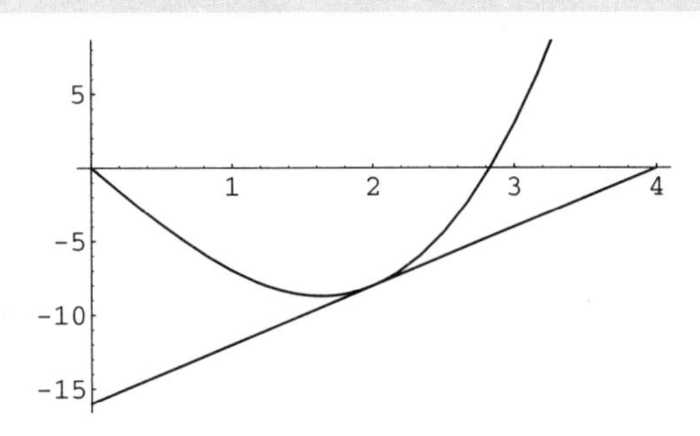

To see that this tangent line closely approximates the graph of the function f near $x = 2$, plot this graph over a small interval about $x = 2$.

```
In[6]:=  Plot[{y, f[x]}, {x, 1.8, 2.2}];
```

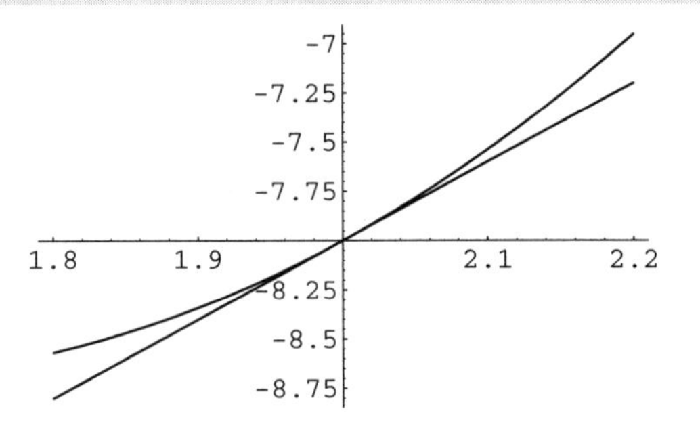

Alternatively we can obtain $f'(2)$ by computing $f'(x)$, and then replacing the variable x with the value 2. We do so by determining the limit shown at the beginning of this section with $a = x$, and assigning the name `Df` to the result.

```
In[7]:=  Df = Limit[(f[x + h] - f[x])/h, h -> 0]
Out[7]=            2
         -8 + 3 x
```

The derivative of f at a general point x is $f'(x) = 3x^2 - 8$. To obtain $f'(2)$, replace x with 2 in `Df` with the command `Df /. x -> 2` (in which `/.` is a shorthand notation for the *Mathematica* command `ReplaceAll`) and obtain the output 4.

Let's try to understand how all the h's disappeared when we took the limit. First, enter the difference quotient $(f(x+h) - f(x))/h$ into *Mathematica*, and store this expression as the variable `diffq`.

```
In[8]:=  diffq = (f[x + h] - f[x])/ h

Out[8]=          3                              3
         8 x - x   - 8 (h + x) + (h + x)
         ---------------------------------
                        h
```

Next, simplify this expression.

```
In[9]:=  Simplify[diffq]

Out[9]=         2              2
         -8 + h   + 3 h x + 3 x
```

Now note that we obtain the derivative, $3x^2 - 8$, by letting h tend to zero (any term containing an h will disappear).

3.2 Differentiating

All the rules for differentiation are programmed into *Mathematica*, making it easy to differentiate complicated expressions. The notation

```
D[expression, x]
```

calculates the derivative of the expression with respect to the variable x.

Example 3.1 Differentiate $3x^2 + 4x^3$ with respect to x.

```
In[10]:=  D[3 x^2+4 x^3, x]

Out[10]=              2
          6 x + 12 x
```

Alternatively, this expression can be assigned the name `f` and then differentiated. Since we assigned a value to `f` earlier in the chapter, we must clear the old definition using the command `Clear`, before using `f` again.

```
In[11]:=  Clear[f]
          f = 3 x^2 + 4 x^3

Out[12]=     2     3
          3 x  + 4 x

In[13]:=  Df = D[f,x]

Out[13]=           2
          6 x + 12 x
```

Here, the derivative is assigned to the variable `Df`. To evaluate the derivative at a point such as $x = 2$, use `/.` to replace all occurrences of x in the expression `Df` with the value 2.

```
In[14]:=  Df /. x -> 2

Out[14]=  60
```

Example 3.2 Enter the expression $\cos^2(t^2)$ in *Mathematica* as follows:

```
In[15]:=  f = Cos[t^2]^2

Out[15]=        2 2
          Cos[t ]
```

Now differentiate f with respect to t.

```
In[16]:=  D[f, t]

Out[16]=              2        2
          -4 t Cos[t ] Sin[t ]
```

You can calculate higher derivatives by specifying the second argument of the form $\{var,\ n\}$, where n designates the number of derivatives that are to be taken. For example, you can compute the second derivative of f by entering the command `D[f,{t, 2}]` (try this).

If f is a function, then $f(x)$ is an expression in x. You can differentiate it by typing `D[f[x], x]` (here, you must type `f[x]` rather than just `f`).

Example 3.3 Enter $f(x) = x^2(x^5 + 1)$ as a function.

```
In[17]:=  f[x_] := x^2 (x^5 + 1)
```

This function can be differentiated with the command

```
In[18]:=  D[f[x], x]

Out[18]=     6           5
          5 x  + 2 x (1 + x )
```

To evaluate $f'(2)$, you can use `/.` by entering

```
In[19]:=  D[f[x], x] /. x -> 2

Out[19]=  452
```

To evaluate the preceding expression, *Mathematica* calculates the derivative of $f(x)$, and then replaces all occurrences of x with 2.

Try evaluating f' at the points $x = -3$, $x = 3$, and $x = t$, by typing the following input.

```
D[f[x], x] /. x -> -3
D[f[x], x] /. x -> 3
D[f[x], x] /. x -> t
```

Example 3.4 Compute the equation of the tangent line at $x = 2$ of the function

```
In[20]:=  f[x_] := (x^3-1)/(x+2)
```

The tangent line passes through the point $(2, f(2))$ and has a slope equal to $f'(2)$, which in *Mathematica* is represented as `f'[2]`. First assign the name `m` to the slope.

```
In[21]:=  m = f'[2]

Out[21]=  41
          --
          16
```

The formula of the tangent line is now given by the expression

```
In[22]:=  y = Expand[m (x - 2) + f[2]]

Out[22]=    27     41 x
          -(--) + ----
            8      16
```

You can verify that this expression is the tangent by graphing it with the function f by invoking `Plot[{f[x], y}, {x, 0, 4}]`.

You can evaluate higher derivatives by repeated application of the `D` operator. For example, `f''[x]`, `D[f[x], x, x]`, or `D[f[x], {x, 2}]` represents the second derivative of the function f (so you can compute $f''(2)$ by entering `f''[2]` or `D[f[x], {x, 2}] /. x -> 2`).

3.3 Summary

- You can use a *Mathematica* function to compute a difference quotient.
- You can compute the slope of a tangent line by taking the `Limit` of a difference quotient.
- You can write a *Mathematica* expression for the line tangent to the graph of $y = f(x)$. You can display both the curve and the line tangent to it on the same graph.
- Given a *Mathematica* function *f* that takes a single argument, you can use `D[`*f*`[`*x*`], `*x*`]` to compute the derivative function with respect to the variable *x*.
- Given a *Mathematica* expression `expr`, you can use `D[`*expr*`, `*x*`]` to calculate an expression for the derivative.
- Given a *Mathematica* expression *expr*, you can use `D[`*expr*`, {`*x*`, `*n*`}]` to calculate higher derivative expressions.

3.4 Exercises

1. Differentiate these expressions in *Mathematica*.

   (a) $\dfrac{t^2 + t}{t^3 - 1}$

   (b) $\cos^2(t^3 + 1)$

2. Assign the expression $r^3 + \sin(r)\cos(r)$ to the variable f. Then, differentiate f with respect to r.

3. Assign the expression $a \sin(x^2) + b \cos(x^2)$ to the variable f. Differentiate f with respect to x (assuming a and b are constants). Now differentiate f with respect to a (assuming x and b constants).

4. For each of exercises 1, 2, and 3, define the given expression as a function; then use the `D` syntax to take its derivative.

5. Define the functions $f(x) = x^3+3x+7$ and $g(x) = \tan(x^2+1)$ in *Mathematica*. Then use *Mathematica* to differentiate $f(x)g(x)$, $f(x)/g(x)$, $(f \circ g)(x)$, and $(g \circ f)(x)$. Here, the composition $(f \circ g)(x)$ can be entered in *Mathematica* as `f[g[x]]`.

6. Find the equation of the tangent line to the function $f(x) = x^4 - x^3$ at the point $x = 1.2$. Plot both the graph of the function and the graph of the tangent line on the same coordinate axes.

7. Find the angle of inclination of the tangent line in exercise 6. *Hint:* Recall that the tangent of the angle of inclination is the slope of the line. So this problem amounts to taking the inverse tangent (`ArcTan`) of the slope of the line in exercise 6.

8. Find the point of intersection of the line $y = 2x + 1$ and the curve $y = \sqrt{24.5 - x^2}$. Then, find the angle between the tangent lines at the point of intersection.

9. *This exercise does not involve Mathematica. It is a warmup for exercise 10, which does involve Mathematica.* Suppose the curve pictured here represents a plot of distance versus time for an object (time is the horizontal axis and distance is the vertical axis). From this plot, determine the approximate time T when the instantaneous velocity of the object equals the average velocity of the object over the interval $0 \leq t \leq T$.

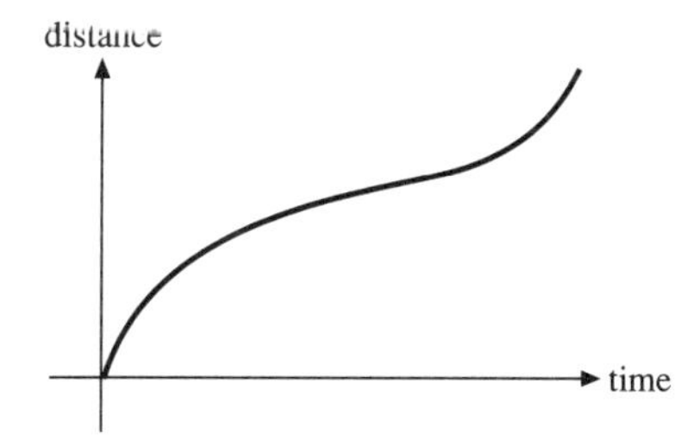

10. Consider the following data, which represent the position (in meters) of an object at various times (in seconds).

Time	0.10	0.20	0.30	0.40	0.50	0.60	0.70	0.80	0.90	1.00
Position	0.25	0.40	0.65	0.84	0.99	1.10	1.20	1.38	1.58	1.80

Plot this data set using *Mathematica*. *Hint:* Enter the data as a list of lists, i.e.,

```
data = {{0.1,0.2},{0.2,0.4}, ... }}
```

as done in Chapter 2. Then type

```
ListPlot[data]
```

This will draw the points. To connect the data points with line segments, enter the command

```
ListPlot[data, PlotJoined -> True]
```

From this plot, estimate the time T at which the instantaneous velocity is equal to the average velocity over the time interval $0 \leq t \leq T$.

11. *Newton's method.* We have already seen examples showing how to solve equations, such as $x^3 + x - 2 = 0$, numerically with the `FindRoot` command. The point of this exercise is to explore one particular algorithm, called Newton's method, that is often used to solve equations numerically. The basic idea behind Newton's method is as follows. To solve the equation $f(x) = 0$, pick a starting point, x_0, near the solution to $f(x) = 0$ (e.g., from the graph, choose the integer closest to the solution). Generally, the x-intercept of the tangent line to $y = f(x)$ at x_0 is closer to the solution than is x_0. We denote this x-intercept by x_1.

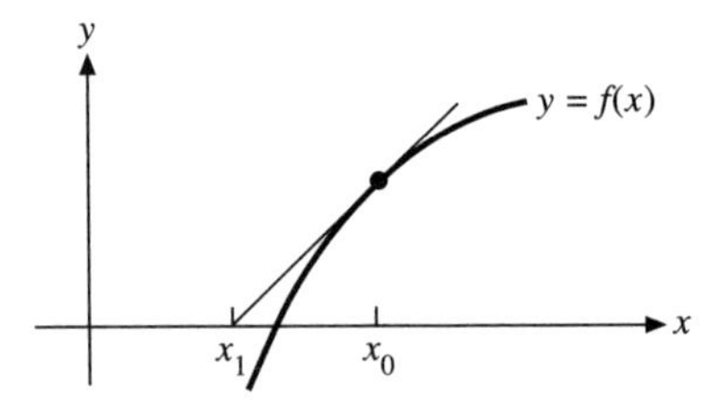

After computing the equation of the tangent line and finding its x-intercept, we obtain:

$$x_1 = x_0 - \frac{f(x_0)}{f'(x_0)}$$

Then, we iterate this process. This iteration procedure leads to a sequence of points:

$$x_{n+1} = x_n - \frac{f(x_n)}{f'(x_n)} \quad \text{for } n \geq 0$$

which approaches the solution to the equation $f(x) = 0$ as n gets bigger (in fact, n usually does not have to get too large).

(a) Program this algorithm into *Mathematica*. Define the function `f[x_] := x^3+x-2` (enter this function definition into *Mathematica*). A plot of the graph of this function reveals that a root to the equation $f(x) = 0$ exists in the interval $0 \leq x \leq 1$.

Define the following function, called `newton`, which yields the next iteration of Newton's method.

```
newton[x_] := N[x - f(x) / D[f[x], x]]
```

Using $x = 0$ as a starting point, `newton[0]` returns the first iteration of Newton's method. Now repeatedly execute the command `newton[%]` until the output no longer changes. (If you are using the Notebook Front End, type <COMMAND> –l (the COMMAND key and the lower-case letter L) or use copy and paste to obtain multiple copies of the command.) Compare your answers with the result of *Mathematica*'s `FindRoot` command.

(b) The following set of commands executes five iterations of Newton's method with $x_0 = 0$.

```
Clear[x]
xNew = 0;
Table[xNew = N[x - f[x]/D[f[x],x]] /. x -> N[xNew],
        {n, 5}]
```

The `Table` command generates a list of the values of the expression `N[x - f[x]/D[f[x],x]] /. x -> xNew` for `n` set equal to 1, 2, 3, 4, and 5. The index variable `n` acts as a counter. The variable `xNew` after this program is executed is set to an approximation to the solution of the equation $f(x) = 0$. Compare this approximation to the solution that you obtained using *Mathematica*'s `FindRoot` command.

(c) A variation of part (b) is to execute Newton's method until the difference between the most recent and next most recent values in the sequence of approximations (i.e., $|x_{n+1} - x_n|$) is less than some preassigned small number, such as 10^{-7}. One way to do this is to keep track of the most recent and next most recent values in the sequence of approximations by the variables x_{new} and x_{old} and then assign the variable $delta_x$ to their difference $|x_{\text{new}} - x_{\text{old}}|$. The algorithm should continue to execute until $delta_x$ is less than the preassigned value. To implement this in *Mathematica*, first assign the value 1 to the variable `deltaX`, and then replace a call to table `Table` with a `While` statement.

```
xNew = 0;
count = 0;
deltaX = 1;
While[ deltaX >= 10^(-7)  && count < 20,
        count = count + 1;
        xOld = xNew;
        xNew = ...;
        deltaX = ...
]
```

Using `While`, repeatedly calculate a new value for x, as long as the variable $delta_x$ exceeds 10^{-7} or until the counter `count` reaches 20. (The choice of 20 is somewhat arbitrary. Its purpose is to keep the program from running indefinitely if $delta_x$ never gets below 10^{-7}.) For this exercise, fill in the *Mathematica* commands in the `While` statement to implement Newton's method on the equation $x^3 + x - 2 = 0$ until $|x_{\text{new}} - x_{\text{old}}|$ is less than 10^{-7}.

12. A cylindrical can with a top and bottom is to contain 1000 cubic centimeters. Find the dimensions of the can if its surface area is 600 square centimeters. *Note:* There are two answers.

13. A sheet metal worker constructs a trough from a 2 foot by 10 foot rectangular piece of metal by bending it so that the 2-foot width forms an arc of a circle (see the figure). If the volume of the trough is 4 cubic feet, find the angle t subtended by the arc.

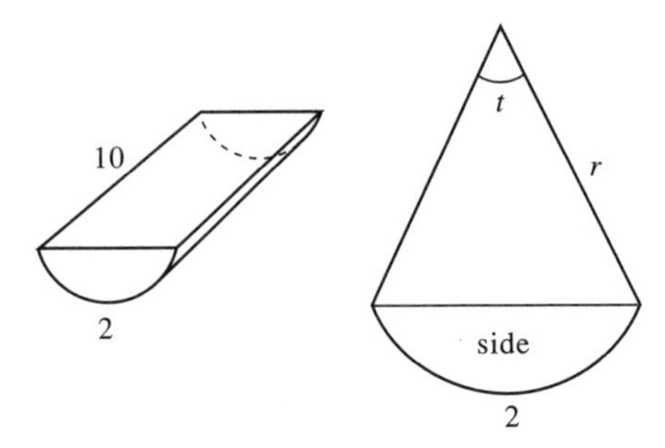

Hint: Suppose that r is the radius of the circular arc subtended by the angle t. Note that $rt = 2$ and that the cross-sectional area of the trough is

$$\tfrac{1}{2}r^2t - r^2 \sin(t/2)\cos(t/2))$$

(Think of the area of the cross-section as the area of the circular arc minus the area of a triangle.) Use these equations to derive the equation of the volume of the trough

$$\textit{volume} = \frac{20(t - \sin(t))}{t^2}$$

Use *Mathematica*'s `FindRoot` command to solve the equation *volume*$= 4$. Use several iterations of Newton's method to solve this equation and compare your answer with the one that you obtained using `FindRoot`.

14. A pulley consists of an 18-inch band tautly wrapped around two wheels of radius 1 inch and 2 inches, respectively, as shown in the diagram. Find the length of the straight pieces of the band that are not in contact with either wheel.

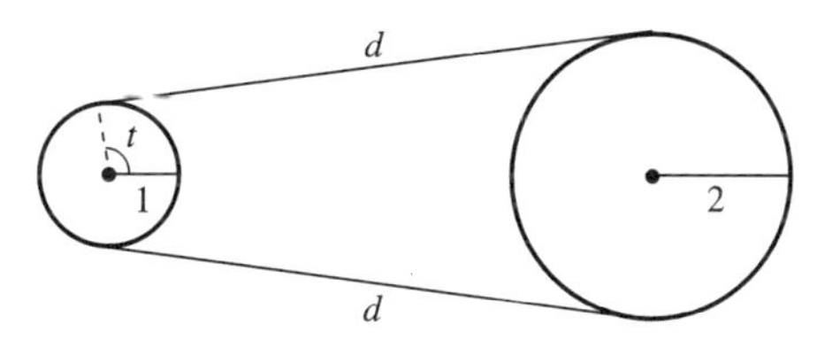

Hint: Let t be the angle of the sector of the smaller wheel (measured from the horizontal) that does *not* make contact with the band. The angle t also represents the angle of the sector of the larger wheel that *does* come in contact with the band (why is this assertion true?). Using the fact that the perimeter of the band is 18 inches, derive the equation

$$2t - 2\pi + 2d = 18$$

Now derive the equation $\tan(t) = d$ (a trigonometry-plus-similar-triangles argument). Solve this equation for d, and substitute it into the first equation. Solve the first equation using `FindRoot`. Use several iterations of Newton's method to solve this equation and compare your answer with the one you obtained using `FindRoot`.

4 Applications of Differentiation

In this chapter, we present *Mathematica* commands for implicitly differentiating an equation involving more than one variable. We also discuss applications to linear approximation and related rates.

4.1 Implicit Differentiation

In previous chapters, we have discussed functions or expressions that are defined explicitly, which means that the dependent variable, such as y or f, appears on one side of an equation, and an expression of x appears on the other side (e.g., $y = x^2$). In this chapter, we consider functions and expressions that are given *implicitly*. In such cases, the variables are mixed together in the equation. An example is the equation $3xy = x^3 + y^3$ that defines the Folium of Descartes. Note that it would be difficult to solve for the variable y explicitly in terms of x. Nevertheless, plots and derivatives can still be obtained.

To plot this equation, we use the function `ImplicitPlot`, which is defined in the package `ImplicitPlot.m` in the directory (or folder) `Graphics`. We read in the package with the `Needs` command, to load the definition of `ImplicitPlot`. Note that the single quotation marks in the argument to `Needs` are back quotation marks.

```
In[1]:=  Needs["Graphics`ImplicitPlot`"];
         Clear[x, y]
         eq = (3 x y == x^3 + y^3)

Out[3]=               3     3
         3 x y == x  + y

In[4]:=  ImplicitPlot[eq, {x, -3, 3}, {y, -3, 3},
                  PlotPoints -> 35];
```

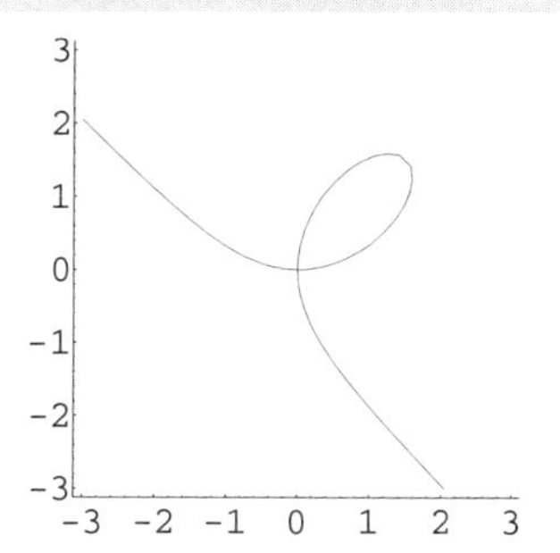

By default, when you call `ImplicitPlot`, *Mathematica* samples the equation at 25 places. If it determines the curve looks sufficiently smooth, it then goes ahead and plots it. If not, it will sample at more points. However, this sampling strategy is not always sufficient to yield an accurate rendering of an implicit curve. Occasionally *Mathematica* is fooled. The remedy for this situation is to increase the minimum number of sample points using the option `PlotPoints`. In the present example, we specify `PlotPoints -> 35` so that `ImplicitPlot` samples the equation at a minimum of 35 places.

Note that this plot contains a loop, which cannot be described globally as the graph of one function $y = y(x)$. Near most points, however, the plot is the graph of one function. For example, the lower piece of the loop over the interval $-1 \leq x \leq 1$ is the graph of one function $y(x)$. Finding a formula for $y(x)$ involves solving the equation $3xy = x^3 + y^3$ for y in terms of x. Using `Solve`, we can obtain an exact solution. However, the solution is messy. Alternatively, using `FindRoot`, we can obtain numerical values of $y(x)$ at specific values of x. For example, here we find the value of y at $x = 1.5$.

```
In[5]:=  x = 1.5;
         eq = (3 x y == x^3 + y^3);
         FindRoot[eq, {y, 1, 2}]

Out[7]=  {y -> 1.5}
```

Therefore, $y = 1.5$ at $x = 1.5$. We specify two starting values, by giving `{y, 1, 2}` as the second argument to `FindRoot`. Note that we obtain the larger of the two possible values of y (see the plot).

A plot over a small range that limits the range of x and y reveals that the plot satisfies the vertical line test near $x = 1.5$.

```
In[8]:=  Clear[x]
         ImplicitPlot[eq, {x,1.25,1.75}, {y,1.25,1.75}];
```

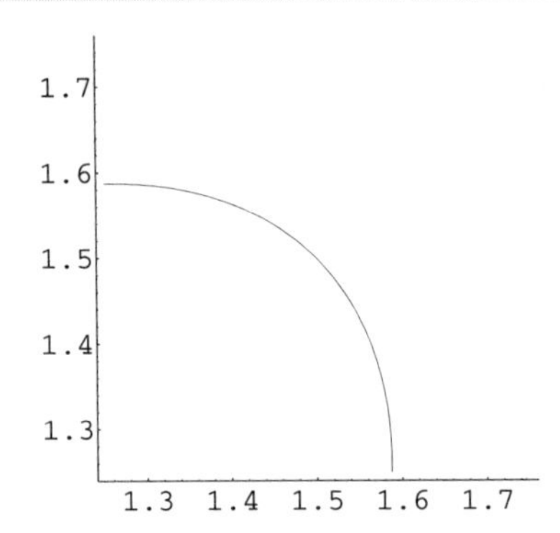

Indeed, over a very small plot range, the graph usually looks like a straight line (the tangent line).

```
In[10]:=  ImplicitPlot[eq, {x,1.49,1.51}, {y,1.49,1.51}];
```

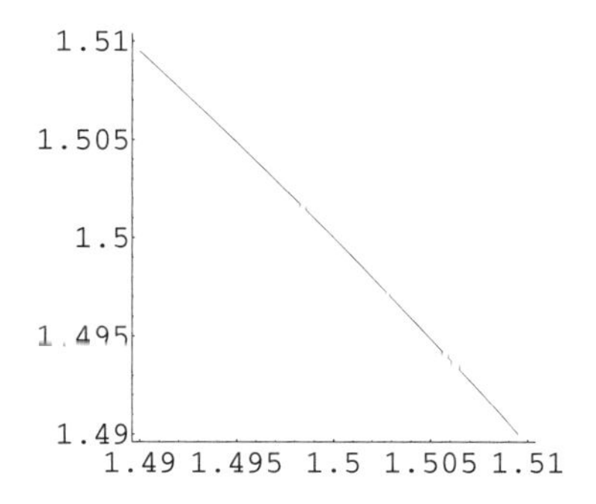

Implicit differentiation is the procedure that we use to find the derivative of an implicitly defined function or expression by using an equation. The following sequence of commands used for implicit differentiation will be applied to the Folium of Descartes, but this sequence of commands applies equally well to other implicitly defined expressions.

First, we assign the name `eq` to the equation. Then, `y` is replaced by `y[x]` (the unknown function)

```
In[11]:=  eq = (3 x y == x^3 + y^3);
          eqNew = eq /. y -> y[x]

Out[12]=                     3        3
          3 x y[x] == x  + y[x]
```

Next, using the command `D`, we take the derivative of both sides of the equation. The symbol `y'[x]` stands for derivative of `y` with respect to `x`.

```
In[13]:=  dEqNew = D[eqNew, x]

Out[13]=                                2         2
          3 y[x] + 3 x y'[x] == 3 x  + 3 y[x]  y'[x]
```

Mathematica uses the chain rule to take the derivative. Now let's isolates y' by using the command `Solve`.

```
In[14]:=  soln = Solve[dEqNew, y'[x]]

Out[14]=                 2
                       -x  + y[x]
          {{y'[x] -> -(----------)}}
                             2
                       x - y[x]
```

After taking the derivative, we no longer need to use the notation `y[x]` to emphasize that `y` depends on `x`. Therefore, we substitute `y` for `y[x]`, and then simplify the derivative and assign it the name `Dy`.

```
In[15]:=  soln /. y[x] -> y

Out[15]=                 2
                       -x  + y[x]
          {{y'[x] -> -(----------)}}
                             2
                       x - y[x]

In[16]:=  Dy = Simplify[%]

Out[16]=              2
                     x  - y
          {{y'[x] -> ------}}
                          2
                     x - y
```

The expression `Dy` is a list with a rule containing a formula or expression for the *implicit derivative*. We obtain a numerical value of the implicit derivative at a point (x, y) by inserting the x-values and y-values into its formula. For instance, the slope at the point $(3/2, 3/2)$ is

```
In[17]:=  m = y'[x] /. Dy[[1]] /. {x -> 3/2, y -> 3/2}

Out[17]=  -1
```

Why did we use the notation `Dy[[1]]`? The expression `Dy` is a list containing a list with the formula. The command `Dy[[1]]` extracts the first element of the list `Dy`. If we had used `Dy`, we would have obtained a list with the solution. The result, -1, agrees with the preceding graph.

4.2 Linear Approximation

The equation of the tangent line of the previous curve can be determined from the information that its slope is -1 and that it passes through the point $(3/2, 3/2)$.

```
In[18]:=  a = 3/2;
          b = 3/2;
          m = -1;
          y = m (x - a) + b

Out[21]=  3 - x
```

The tangent line represents the linear approximation to the graph of the original function. We define a function called `linearApprox` using the tangent line.

```
In[22]:=  linearApprox[x_] := Evaluate[y]
```

The function `linearApprox` approximates the values of y for x near 1.5. To plot both the curve and its tangent line, we use the `Show` command described in Chapter 2.

```
In[23]:=  Clear[x, y]
          p1 = Plot[linearApprox[x], {x, 1, 2},
                 AspectRatio -> Automatic];
          eq = (3 x y == x^3 + y^3);
          p2 = ImplicitPlot[eq, {x, 1, 2}, {y, 1, 2}];
          Show[p1, p2, AspectRatio -> Automatic];
```

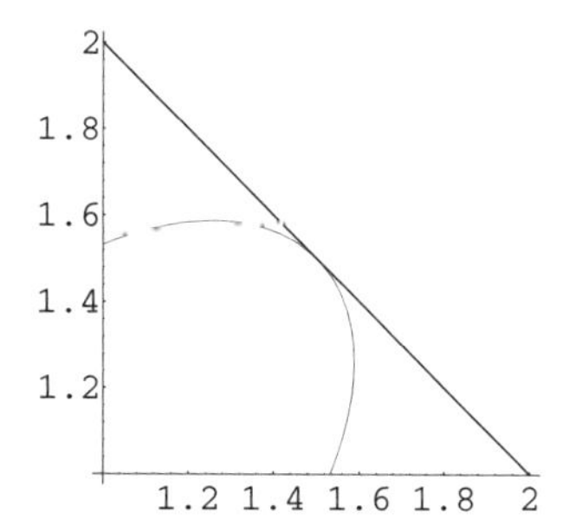

Let's compare the value of `y` and `linearApprox[x]` at $x = 1.51$. To find the value of y at $x = 1.51$, we use `/.` with the rule `x -> 1.51`, and then `FindRoot`.

```
In[28]:=  eq1 = eq /. x -> 1.51

Out[28]=                           3
          4.53 y == 3.44295 + y

In[29]:=  FindRoot[eq1, {y, 1.4, 1.6}]

Out[29]=  {y -> 1.48944}
```

On the other hand, `linearApprox[1.51] = 1.49` as you can easily check. This value is close to the actual value of $y = 1.489436$ given previously.

4.3 Related Rates

In many situations, two or more quantities are related by some formula and are changing with time. In a related-rates problem, the goal is to determine how fast one of the quantities is changing when the rates of change of the other quantities are known.

The procedure to solve a related-rates problem for the case of two related variables is as follows:

1. Determine the two quantities that are changing with time.
2. Find an equation that relates the two quantities (this may involve known formulae from physics or geometry).
3. Enter the equation into *Mathematica*, replacing the two quantities, say u and v, by $u(t)$ and $v(t)$, to reflect that the quantities change with time.
4. Take the implicit derivative of the equation with respect to time t.
5. Solve the resulting equation for the desired derivative.
6. Find numeric values for each of the quantities in the resulting solution, possibly by using the equation in step 3, and substitute them into the result.

The following is a typical example.

Example 4.1 For a concave lens with focal length 30 centimeters, the formula from optics $1/30 = 1/u + 1/v$ expresses the relationship between the distance u of an object from the lens and the distance v of its image from the lens. Suppose that an object is moving toward the lens at a rate of 3 centimeters/second. Find the rate at which the image is receding from the lens when the object is 90 cm away.
Solution. The two quantities that change with time are the two distances, u and v. The equation that relates them is the optics formula given in the problem. This equation is entered into *Mathematica* with $u(t)$ and $v(t)$, signifying that the distances are time dependent.

```
In[30]:=  1/30 == 1/u[t] + 1/v[t]

Out[30]=  1        1       1
          -- == ---- + ----
          30    u[t]   v[t]
```

Now we differentiate the equation with respect to t.

```
In[31]:= D[%, t]

Out[31]=          u'[t]     v'[t]
         0 == -(-----) - -----
                     2        2
                 u[t]     v[t]
```

Now we can solve the resulting equation for $v'(t)$.

```
In[32]:= vRate = Solve[%, v'[t]]

Out[32]=                      2
                          v[t]  u'[t]
         {{v'[t] -> -(-----------)}}
                                2
                            u[t]
```

Now let's calculate the numerical value of v.

```
In[33]:= 1/30 == 1/90 + 1/v

Out[33]= 1     1    1
         -- == -- + -
         30    90   v

In[34]:= imageLength = Solve[%, v]

Out[34]= {{v -> 45}}
```

We find the answer by substituting the values of $u'(t) = 3$, $u(t) = 90$, and $v(t) =$ `imageLength` into the variable `vrate`.

```
In[35]:= imageRate = vRate /.
             {u'[t] -> 3, v[t] -> imageLength, u[t] -> 90}

Out[35]=                 3
         {{v'[t] ->  (-)}}
                         4
```

4.4 Summary

Implicit Differentiation

- You can use `D[eq, x]` to take the derivative of a *Mathematica* equation.

- *Mathematica* expressions and *Mathematica* functions both correspond to the idea of an explicit mathematical function, $y = f(x)$, where there is an independent variable x and a dependent variable y. A vertical line crosses the graph at most once for a given x, and there is an explicit rule to compute y given x.
- You can use `ImplicitPlot`, defined in the package `ImplicitPlot.m` in the directory (or folder) `Graphics.m`, to plot a *Mathematica* equation.
- At points where a vertical line crosses the graph more than once, you cannot solve the equation to get a single function.
- Regardless of whether *Mathematica* can solve an equation explicitly to isolate y, if the vertical line test works in a box, we say that y is defined implicitly as a mathematical function of x, i.e., $y = y(x)$ for some possibly unknown formula $y(x)$. The implicit derivative is the derivative of that (possibly unknown) formula. *Mathematica* can find this derivative, even if it cannot solve for y.
- You can compute the implicit derivative. The text enclosed between the symbols `(*` and `*)` are comments for your edification. They are ignored by *Mathematica*.

```
eqNew = eq /. y -> y[x];
                 (* View y as an implicit function of x. *)
eqNew2 = D[eqNew, x];
                 (* Differentiate both sides of x.       *)
soln = Solve[eqNew2, y'[x]];
                 (* Solve for y'[x].                     *)
deriv = y'[x] /. soln[[1]] /. y[x] -> y;
                 (* Suppress the dependence on x         *)
Dy = Simplify[deriv]
                 (* Simplify the derivative              *)
```

Linear Approximation

- The linear approximation to a curve at a point (x_0, y_0) on the curve is the *Mathematica* function given by the formula for the tangent line.

```
m = Dy /. {x -> x0, y -> y0}
        (* Insert coordinates for point in derivative. *)
y = m (x - x0) + y0
        (* Point-slope formula for tangent line.       *)
linearApprox[x_] := y
        (* Convert the expression to a function.       *)
```

Related Rates

- In a related-rates problem, you suppose that two quantities related by a *Mathematica* equation, *eq*, are both changing with time. Given the rate at which one changes, you are to find the rate at which the other changes.
- The procedure is as follows. You read the problem, identify quantities that change, and produce an equation that relates the two quantities, possibly using known formulae from geometry, physics, or the like.
- For example, suppose that the quantities are denoted by u and v, and the equation relating them, *eq*, has been entered into *Mathematica*. The known rate is $u'_{0,}$ and u_0 is the value of u.

```
eqNew = eq /. {u -> u[t], v -> v[t]}
             (* u & v are implicit functions of time. *)
diffEqNew = D[eqNew, t]
             (* Take the implicit derivative.         *)
vRate = Solve[diffEqNew, v'[t]][[1]]
             (* Solve for the derivative.             *)
vSol = Solve[eq /. u -> u0, v][[1]]
             (* Substitute for u, and solve for v.    *)
Dv = v'[t] /. vRate //.
        {u'[t] -> uPrime0,          (* Plug in numbers. *)
         u[t] -> u0, v[t] -> v, vSol[[1]]}
```

4.5 Exercises

1. By following the steps given here, show that the derivative that you obtain implicitly (without solving for y) is the same as the derivative that you obtain explicitly by solving for y.

   (a) Consider the equation $x^2 + y^2 = 1$ Compute the implicit derivative and assign it to Dy.

   (b) Use *Mathematica* to solve the equation for y and assign the result to `sol`.

   (c) Use the *Mathematica* command `Dy /. sol[[1]]` (and `Dy /. sol[[2]]`) to replace y by its explicit version.

   (d) Use the *Mathematica* command `D[sol[[1]], x]` to take the derivative explicitly.

   (e) Compare your answers in parts (c) and (d).

(f) Repeat steps (c) through (e) with `sol[2]`.

2. One of the virtues of the implicit derivative process is that, given an equation relating x and y, it is not necessary to solve for y explicitly to compute y'. On the other hand, the answer is sometimes difficult to intuit.

   (a) Consider the equation $y^2 - 3xy + 2x^2 = 0$. Use the procedure described in the text to compute the implicit derivative.

   (b) Now load the package `ImplicitPlot.m` by invoking the command `Needs["Graphics`ImplicitPlot`"]` and then call `ImplicitPlot` to draw the graph. Obtain an answer for the derivative that is clearer than the one you got in part (a).

   (c) Try `Factor[eq]`. Now explain why the answer in part (b) is true.

   (d) Without executing a series of *Mathematica* commands, use the insight that you have gained from *Mathematica* graphs and `Factor` to give the linear approximation $L(x)$ at any point (a, b) on the graph, except the origin. Can you find the linear approximation at the origin?

   (e) In the case of the Folium of Descartes, $3xy = x^3 + y^3$, for *most* points (x, y) on the graph there is a plot range for which the restricted graph passes the vertical line test. However, there are two exceptions. Using the `ImplicitPlot` command, click with your mouse to find approximate floating-point decimal coordinates for the two points on the graph that do not have this property.

   (f) Refine your guess about the coordinates of the two points in part (e) by noting that, to solve for y', you have to divide by $x - y^2$. This division could fail if $x - y^2 = 0$ for a point (x, y) on the curve. Use `NSolve[{x - y^2 == 0, eq}, {x, y}]`.

   (g) Use `Solve[{x - y^2 == 0, eq}, {x, y}]` to find the *exact* x and y coordinates of the two points.Confirm that your exact value is correct by using the `N` command on the exact answer to see whether it matches the `NSolve` answer.

3. For each of the following, use `ImplicitPlot` to plot the given equation. Then, find the y-values of the lower piece of the plot at $x = 1, 1.25, 1.5, 1.75, 2$. Compute the slope of the tangent line at $x = 1$ by implicit differentiation. Compare this value to the slope of the chord between $x = 1$ and $x = 1.25$.

   (a) $(x^2 + y^2)^2 = 16(x^2 - y^2)$

   (b) $x^2y + xy^2 = 16$

4. Graph the equations $y = x^2$ and $x^4 + 4x + 2y^2 + 6y = 12$. Find the point(s) of intersection of these graphs. For each point of intersection, find the acute angle between the tangent lines of both equations.

5. When a ray of light hits the surface of a lake, the beam is bent. The equation that governs this effect was discovered by Willebrod Snell (1591–1626). He noted that

$$\frac{\sin(\theta_1)}{\sin(\theta_2)} = 1.33$$

where 1.33 is the index of refraction of water (relative to air), and θ_1 and θ_2 are the angle of incidence and angle of refraction, respectively, measured from a line perpendicular to the surface of the lake. If a fisherman sees the angle of incidence decreasing at $\pi/24$ radians/hr when $\theta_1 = \pi/3$, how fast do fish see the sun rise?

5 Graphs

In this chapter, we use *Mathematica* commands that we introduced earlier, to study graphs of functions and their derivatives.

5.1 Local Maxima and Minima

Recall that, if f is a differentiable function on an open interval, then its derivative is zero at each local maximum or minimum. We can find the local maxima and minima by first graphing the function to get an approximate idea of the location of the maxima and minima and then using `Solve`, `NSolve`, or `FindRoot` to find the solutions of the equation $f' = 0$. `Solve` yields all symbolic solutions of polynomial equations up to fourth order. `NSolve` yields all numerical solutions of polynomial equations of any order. `FindRoot` finds a single numerical root of any type of equation.

Example 5.1 Plot the expression $x^3 + 0.2x^2 - x$, and find the location of the local maxima and minima.

Solution. Assign the name `df` to the derivative of the expression $x^3 + 0.2x^2 - x$.

```
In[1]:=  Clear[f]
         f = x^3 + 0.2 x^2 - x;
         df = D[f, x]

Out[3]=                    2
         -1 + 0.4 x + 3 x
```

Use `Plot[f,{x, -2, 2}]` to observe that the graph of `f` has a local maximum between -1 and 0 and a local minimum between 0 and 1. To find the precise location of the local maximum and minimum, we must solve the equation $f'(x) = 0$ for x. Let us assign the name `xMax` to the local maximum that is between -1 and 0:

```
In[4]:=  xMax = FindRoot[df == 0, {x, -0.5}]

Out[4]=  {x -> -0.647853}
```

and the name `xMin` to the local minimum that is between 0 and 1.

```
In[5]:=  xMin = FindRoot[df == 0, {x, 0.5}]

Out[5]=  {x -> 0.51452}
```

Inserting these values back into the original expression $x^3 + 0.2x^2 - x$ will give the corresponding y-coordinates of the local maximum and local minimum.

```
In[6]:=  f /. {xMax, xMin}

Out[6]=  {0.459883, -0.325365}
```

Therefore, the local maximum is the point $(-0.647853, 0.459883)$ and the local minimum is $(0.51452, -0.325365)$.

You could have solved the previous example without *Mathematica*. However, the next example and the exercises are too complicated for many people to do without the aid of a computer.

5.2 Graphical Analysis

Now let's consider the function:

$$f(x) = \frac{e^x}{x^3 + x - 1 + 0.2e^x}$$

Enter this function into *Mathematica* with e^x entered as `E^x`. Our goal is to graph the function and locate the horizontal and vertical asymptotes, local maxima and minima, and inflection points.

```
In[7]:=  Clear[f]
         f[x_] := Exp[x]/(x^3 + x -1 + 0.2 Exp[x])
         Plot[f[x], {x,-5,15}, AxesLabel->{"x", "f[x]"}];
```

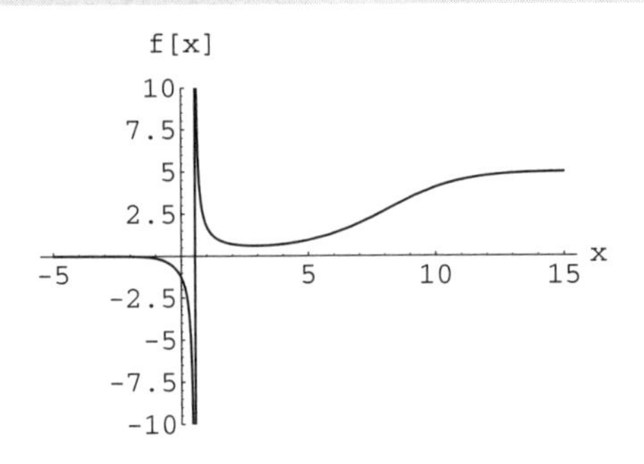

From the graph, it appears that there is a vertical asymptote between $x = 0$ and $x = 1$. We can pinpoint its location by determining the root of the denominator of f. In *Mathematica*, we can obtain the denominator of a function $f(x)$ by using the command `Denominator`. Here, we call `Evaluate` to cause *Mathematica* to evaluate `Denominator[f[x]]` and to obtain an expression, which `FindRoot` can then evaluate.

```
In[10]:=  FindRoot[Evaluate[Denominator[f[x]] == 0],
              {x, 0.5, 0, 1}]

Out[10]=  {x -> 0.521389
```

The lines $y = 0$ and $y = 5$ appear to be horizontal asymptotes as $x \to -\infty$ and $x \to \infty$, respectively. We can verify that they are by taking limits (note that, when x is a large negative number, $f(x) \approx 0$ and, when x is a large positive number, $f(x) \approx \dfrac{e^x}{0.2e^x} = 5$). The command `Apart` rewrites a rational expression as a sum of terms with minimal denominators. In other words, it returns the partial fractions of a rational expression.

```
In[11]:=  Limit[f[x], x -> -Infinity]

Out[11]=  0

In[12]:=  Limit[Apart[f[x]], x -> Infinity]

Out[12]=  5.
```

From the plot, there appears to be a local minimum between $x = 2$ and $x = 4$. We can pinpoint its location by taking the derivative and locating its root.

```
In[13]:=  df = D[f[x], x];
          xMin = FindRoot[df == 0, {x, 3}]

Out[13]=  {x ->  2.893289}
```

The value of the expression `f[x]` at the local minimum is 0.607343, as we can determine by using the command `f[x] /. xMin`, which computes the value of `f[x]` at the value of `x` returned by `FindRoot`.

Let's now make a graph of `f[x]` and its derivative over the interval $0.5 \leq x \leq 15$, because the local minimum of `f[x]` is located there.

```
In[14]:=  Plot[{f[x], df}, {x, 0.5, 15},
              PlotStyle -> {GrayLevel[0], Dashing[{0.02}]},
              AxesLabel -> {"x", ""}]
```

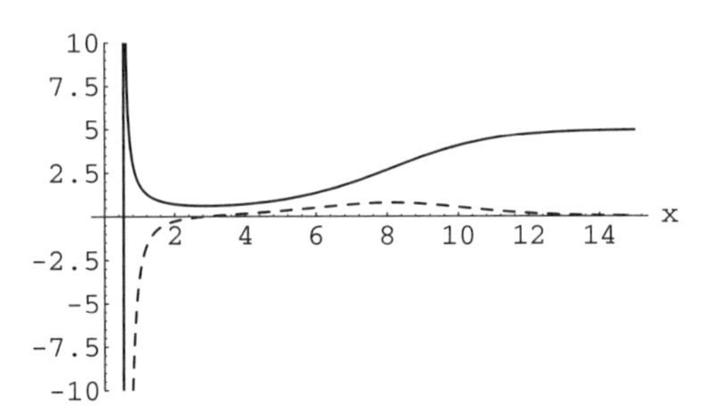

Note that f decreases on the interval $0.5 < x < 2.89$, and the derivative of f is negative on this same interval. Likewise, f increases on $x > 2.89$, and the derivative of f is positive on this same interval. The graph of f flattens as x gets large. This corresponds to the observation that the graph of the derivative approaches $y = 0$ when x gets large.

There also appears to be an inflection point (a point where the concavity of the graph switches) between $x = 6$ and $x = 10$. We can pinpoint its location by taking the second derivative (the derivative of the first derivative, `df`) and finding its root.

```
In[15]:=  ddf = D[df, x];
          inflectionPt = FindRoot[ddf == 0, {x, 8, 6, 10}]

Out[16]=  {x -> 8.1314}

In[17]:=  f[x] /. inflectionPt

Out[17]=  2.77585
```

The value of the expression f at $x = inflectionPt$ is 2.77585. So the point of inflection is approximately $(8.131, 2.776)$.

We now plot the expression f and its second derivative on the same coordinate axis.

```
In[18]:=  Plot[{f[x], ddf}, {x, 0.5, 15},
              PlotRange -> {-1, 5},
              PlotStyle -> {GrayLevel[0], Dashing[{0.02}]},
              AxesLabel -> {"x", "y"}
          ];
```

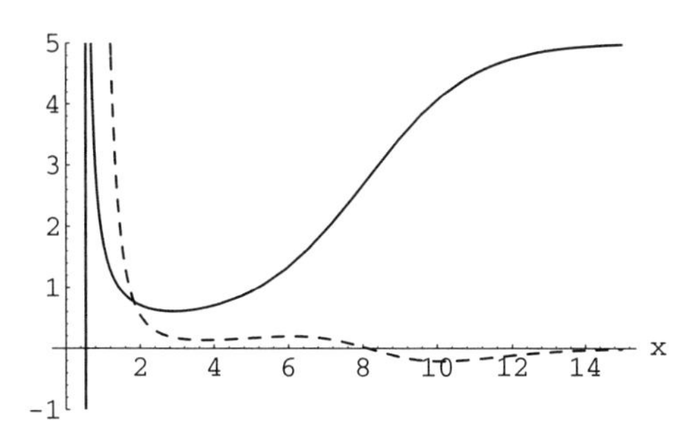

Note that we changed the range of y-values displayed by using the option `PlotRange -> {-1,5}`. The function f is concave up and the second derivative is positive on the interval $0.5 < x < 8.13$. The graph of f is concave down and the second derivative is negative on the interval $x > 8.13$. Also note that the graph of f is steepest where the first derivative is largest (see previous graph). This point, $x = 8.13$, is where the second derivative is zero.

5.3 Designer Polynomials

Example 5.2 Here is an example that combines the use of many of the *Mathematica* commands introduced up to this point. The problem is to find the coefficients of the cubic polynomial $f(x) = ax^3 + bx^2 + cx + d$ that has a local minimum at $(-2, 0)$ and a local maximum at $(3, 4)$.

Solution. Four equations are needed for the four unknowns *a, b, c,* and *d.* We will assign the names `eq1` through `eq4`, so that we may refer to them easily later. After defining f as a *Mathematica* function, we can obtain the first two equations from the information $f(-2) = 0$ and $f(3) = 4$.

```
In[19]:=  Clear[f, a, b, c, d, x]
          f[x_] := a x^3 + b x^2 + c x + d;
          eq1 = (f[-2] == 0);
          eq2 = (f[3] == 4);
```

The remaining two equations can be obtained from the fact that the derivative of f must be equal to zero at $x = -2$ and $x = 3$ (because f has a local minimum and maximum at these points).

```
In[23]:=  eq3 = (f'[x] == 0) /. x -> -2;
          eq4 = (f'[x] == 0) /. x -> 3;
```

Now these four equations are solved for *a, b, c,* and *d.*

```
In[25]:=  solution = Solve[{eq1,eq2,eq3,eq4}, {a,b,c,d}]

Out[25]=          176              8             12            144
          {{d -> ---, a -> -(---), b -> ---, c -> ---}}
                  125             125            125           125
```

We assign them the name `solution` so we can easily refer to this solution. Note the use of the curly braces { } to enclose `eq1, eq2, eq3, eq4` and `a, b, c, d`. The first argument to `Solve` is a list of the equations to be solved, and the second argument is a list of the variables for which solutions are sought. If you have only a single equation or variable, it needs not be enclosed in a list. The output `solution` is given in the form of a list of replacement rules. We can obtain the value of `a` in the solution by typing `a /. solution[[1]]`. Similarly, we can obtain the value of `b` by typing `b /. solution[[1]]`. To substitute values of *a, b, c, d* back into the original function, type

```
In[26]:=  f[x] /. solution[[1]]

Out[26]=                           2        3
          176     144 x     12 x      8 x
          --- + ----- + ----- - ----
          125     125       125      125
```

Plot the graph of this solution to see whether it has a local minimum at $(-2, 0)$ and a local maximum at $(3, 4)$ (do the plot!).

5.4 Summary

We have shown you how to use *Mathematica*'s `Plot` and `Solve` commands in concert to find intercepts, local maxima and minima, horizontal and vertical asymptotes and inflection points. Make sure that you are familiar with the following techniques.

- To find x-intercepts, plot the expression. Then call `FindRoot` specifying a starting value, s_{start}, where the curve is approximately 0 on the graph.

```
Plot[expr, {x, xmin, xmax}]
FindRoot[expr == 0, {x, xstart}]
```

- To find local maxima and minima, plot the expression. Compute the derivative. Call `FindRoot` with the approximate x-coordinates of local maxima and minima in your graph. Then use `/.` (`ReplaceAll`) to find the y-coordinate of the critical points.

```
Plot[expr, {x, xmin, xmax}]
df = D[expr, x]
ans = FindRoot[df == 0, {x, xstart}]
f[x] /. ans
```

- To find inflection points, plot the function. Compute the second derivative. Find the root and then determine the y-coordinate of the inflection point.

```
Plot[f[x], {x, xmin, xmax}] (* Plot function.              *)
ddf = D[f, {x, 2}]    (* Compute the second derivative.*)
ans = FindRoot[ddf == 0,
         {x, xStart, xLowerBound, xUpperBound}]
        (* Restrict search, approximate x-coordinate. *)
f[x] /. ans   (* Find y coordinate of inflection point.*)
```

- To find vertical asymptotes, plot the function. Isolate the denominator. Find the root of the denominator.

```
Plot[f[x], {x, xmin, xmax}]}  (* Plot the function.    *)
Together[f[x]]   (* Make function into single fraction.*)
Denominator[f[x]]   (* Isolate denominator.            *)
FindRoot[Evaluate[Denominator[f[x]] == 0],
    {x, xStart, xLowerBound, xUpperBound}]
  (* Restrict search, approximate zero of denominator. *)
```

- To find horizontal asymptotes:

```
Limit[f[x], x -> Infinity]     (* Find limit on right. *)
Limit[f[x], x -> -Infinity]    (* Find limit on left.  *)
```

5.5 Exercises

1. Plot the expression $f = 0.25x^4 - 0.913x^3 + x^2 - 0.175x - 0.5$. You may need to adjust the plot ranges to include all relevant aspects of the graph. Find the local maximum (or maxima) and minimum (or minima) and the intervals where f is increasing and decreasing.

2. Find the cubic $ax^3 + bx^2 + cx + d$ that has a local maximum at $(-1, 2)$ and a local minimum at $(3, -2)$.

3. Find the cubic polynomial that passes through the points $(-1, -2)$, $(1, 3)$, and $(4, 2)$, and whose derivative at $x = 1$ is 1.7.

4. Find the cubic polynomial with a local maximum at $(-2, 4)$ and a local minimum at $(3, -1)$.

5. Plot an informative graph of

$$y = \frac{x^3 - 6x^2 + 11x - 5.3}{x^3 - 3.1x^2 - 3.2x + 4.21}$$

Determine all asymptotes and critical points.

6. Graph the function

$$y = \frac{x \ln(x)}{x^2 + 1}, \text{for} x > 0$$

Find all critical points and inflection points. Find intervals where the function is increasing, decreasing, concave up, and concave down.

7. For each of the following expressions, locate:
   (i) the horizontal and vertical asymptotes

   (ii) the local maxima and minima

   (iii) the inflection points.

   Then, plot the given expression, along with its derivative and second derivative. Observe that the intervals where the expression is increasing (decreasing) correspond to the intervals where its derivative is positive (negative). Do a similar analysis comparing concavity for the expression and the sign of the second derivative.

   (a) $\dfrac{10x^2 - x + 1}{10x - 1}$

   (b) $\dfrac{3x^5 + 2x}{x^5 - 3x + 1}$

   (c) $\dfrac{2e^x}{e^x - x^3}$

   (d) $x^2 + 5\cos(x)$

   (e) $\ln(x) - 4x^2 + x^3$

8. Each of the following plots represents the graph of the *derivative* of a function. Draw a plausible graph of the corresponding function, and locate (approximately) any local maxima or minima and inflection points.

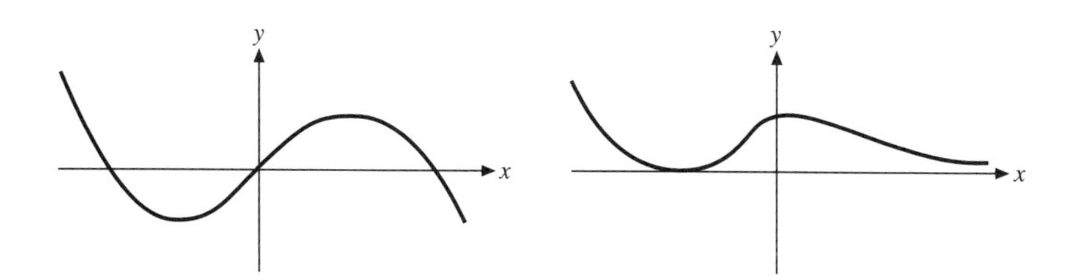

9. Each of the following triplets of plots represents the graph of a function, its derivative, and its second derivative (although not necessarily in that order). Determine which graph represents the function, the derivative, and the second derivative.

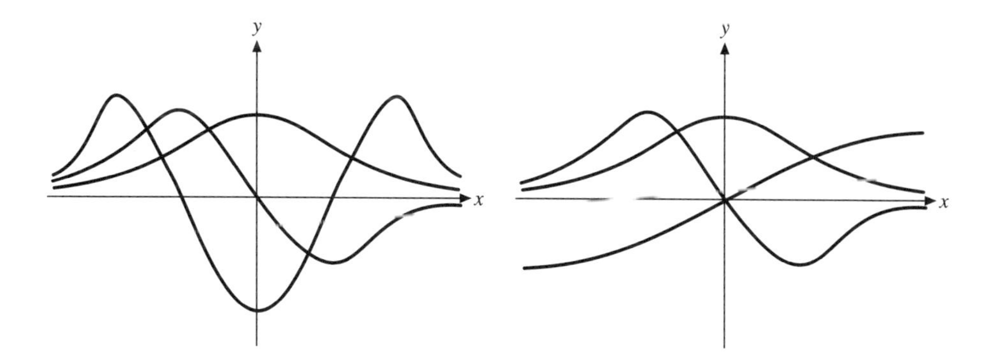

10. (This exercise is a warmup for exercise 11) A plot of land has a flat, horizontal southern boundary and an irregular northern boundary. The goal of this problem is to draw a map of the boundary of this plot of land. Since the southern boundary is horizontal, it will be represented by the x-axis with the origin located at its western point. North-south measurements (represented by the variable y) are taken, and are placed in the following table (each unit represents 100 feet).

x-value	0.0	1.0	2.0	3.0	4.0
y-value	0.0	2.1	3.2	2.4	1.7

The first task is to find the graph of the quadratic polynomial p (a parabola) that contains the first three data points. You can find it by defining the quadratic polynomial $p(x) = ax^2 + bx + c$ as a function and then solving the equations $p(0) = 0$, $p(1) = 2.1$, and $p(2) = 3.2$ for the unknowns a, b, and c. The next task is to find the graph of the cubic polynomial q that contains the last three data points and is such that $p'(2) = q'(2)$. This last requirement ensures that the slope of the graph of p at $x = 2$ will agree with that of q at $x = 2$. To find the graph, enter $q(x) = dx^3 + ex^2 + fx + g$ as a function (if necessary, unassign any previously used labels). Enter the equations $q(2) = 3.2$, $q(3) = 2.4$, $q(4) = 1.7$, and $p'(2) = q'(2)$; then solve for the unknowns d, e, f, and g.

Graph p and q on the same plot. As explained in Chapter 2, to display two graphs together enter the commands `g1 = Plot[p[x], {x,0,2}]; g2 = Plot[q[x], {x,2,4}]`. Then issue the command `Show[g1, g2]`. This sequence of commands is necessary because p and q are plotted over different intervals.

11. *Smooth off the southern boundary of Texas.* In exercise 25 in Chapter 2, the boundary of the state of Texas was drawn using line segments. The point of this problem is to use parabolas and cubics to smooth out the part of the southern boundary of the state formed by the Rio Grande. Enter the relevant Rio Grande data:

```
rio = {{0,0},{1,-1.1},{2,-2.5},{3,-2.9},{4,-2.3},
          {5,-2.8},{6,-4.4},{7,-5.8},{8,-6.1}};
```

Here the origin is the western corner of Texas (near El Paso), and the x-axis is the extension of the east-west border between New Mexico and Texas. Each unit represents approximately 69 miles.

To find functions that smooth out the Rio Grande, proceed as in exercise 10. First, find the parabola $p(x)$ that passes through the three data points $(0, 0)$, $(1, -1.1)$, and $(2, -2.5)$. Then, find a cubic polynomial q that passes through the next triplet of data points $(2, -2.5)$, $(3, -2.9)$, $(4, -2.3)$, and that also satisfies the equation $p'(2) = q'(2)$ (so that the slopes of the graphs of p and q at $x = 2$ are the same). In the same manner, find cubics r and s for the triplets $(4, -2.3)$, $(5, -2.8)$, $(6, -4.4)$ and $(6, -4.4)$, $(7, -5.8)$, $(8, -6.1)$ so that $q'(4) = r'(4)$ and $r'(6) = s'(6)$. Then, plot the following on the same coordinate axes: $p(x)$ over $0 \le x \le 2$; $q(x)$ over $2 \le x \le 4$; $r(x)$ over $4 \le x \le 6$; and $s(x)$ over $6 \le x \le 8$. Use the `Show` command as in exercise 10.

6 Applied Max/Min

The first section of this chapter describes how to find the absolute maximum or minimum of a function on an interval. The second section presents an example of an applied optimization problem that would be difficult to solve without the aid of a computer.

6.1 Absolute Extremes of a Function on an Interval

In this section, we use *Mathematica* to help find the absolute maximum or minimum of an expression or a function on an interval. The method compares the values of the function or expression at its critical points on the given interval with those at the endpoints of the interval. This approach is illustrated by the following example.

Example 6.1 Find the absolute maximum and minimum of the function $f = \ln(x) - 4x^2 + x^3$ on the interval $1 \leq x \leq 3$.

Solution. First, define f as *Mathematica* function. Then, plot it over the interval $1 \leq x \leq 3$.

```
In[1]:=  Clear[f]
         f[x_] := Log[x] - 4x^2 + x^3
         Plot[f[x], {x, 1, 3}];
```

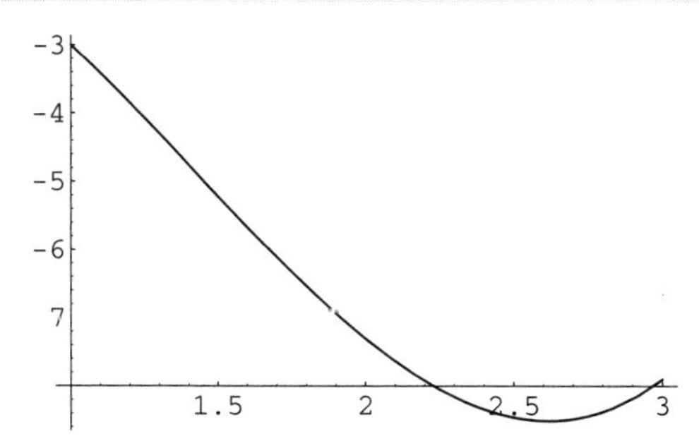

From the plot, it appears that the minimum of f is located at a critical point near $x = 2.5$, and that the maximum of f is located at the left-hand endpoint $x = 1$. If you are using the Notebook Front End version of *Mathematica*, you can see in the lower edge of a window the approximate coordinates of a point by clicking the mouse button on a graph, then holding down the $\langle$COMMAND$\rangle$ key and positioning the pointer above the point on the graph. To find the critical point, set the derivative of f equal to zero, and solve. Here we specify the starting value of $x = 2$ to FindRoot.

```
In[4]:=  df = f'[x]

Out[4]=  1            2
         - - 8 x + 3 x
         x

In[5]:=  xMin = FindRoot[f'[x] == 0, {x, 2}]

Out[5]=  {x -> 2.61803}
```

The values for f at the endpoints $x = 1$ and $x = 3$, as well as those at $x = xMin$, are approximately -3, -7.90139, and -8.50971, respectively.

```
In[6]:=  {f[1.], f[3.], f[x] /. xMin}

Out[6]=  {-3., -7.90139, -8.50971}
```

So, the maximum of f is -3 at $x = 1$, and the minimum of f is -8.50971 at $x = 2.61803$.

If the interval is open (does not contain its endpoints) or if the interval is infinite in length, then there may not be an absolute maximum or minimum. As an example, consider a plot of the function

$$f = \frac{x^4 + 2x + 3}{x^2}$$

over the interval $x > 0$.

```
In[7]:=  Clear[f]
         f[x_] := (x^4 + 2x + 3)/x^2
         Plot[f[x], {x, 0, 20}, PlotRange -> {-100, 100}];
```

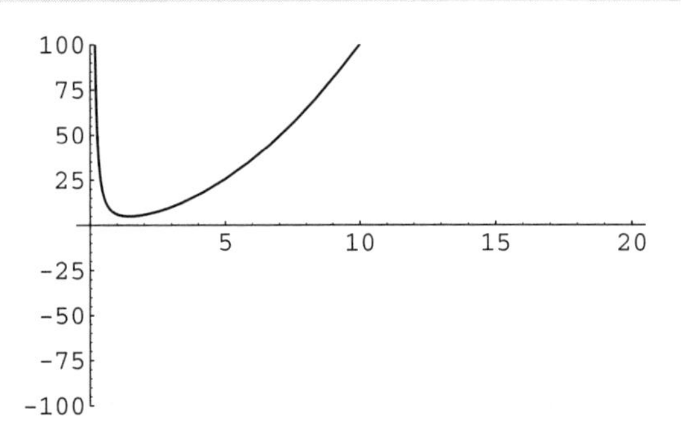

The plot reveals an absolute minimum at a critical point near $x = 1.5$. However, this expression has no absolute maximum, since f approaches infinity as $x \to 0$ or $x \to \infty$. We can find the minimum by computing the derivative, `df = f'[x]`, and then setting the derivative equal to zero, `FindRoot[df == 0, {x, 1}]`. The minimum of f is approximately 4.90866, occurring at $x = 1.45263$.

6.2 The Most Economical Tin Can

For an applied max/min problem, you must first translate the problem into mathematics (*Mathematica* will not help you here), and then solve the resulting calculus problem (which is where *Mathematica* can help).

Example 6.2 You are to construct a cylindrical tin can by joining the ends of a rectangular piece of material to form the cylindrical side and then attaching circular pieces to form the top and bottom. There are seams around the perimeter of the top and bottom, and there is one seam down the side surface (where the ends of the rectangle join together). Suppose that the volume of the can is 1000 cubic centimeters. Also suppose that the cost of the material is $1.00 per square meter and the cost of the seam is $0.20 per meter. Find the dimensions of the can that will minimize its cost.

Solution. Let `r` be the base radius of the can and `h` be the height of the can. The surface area of the can is

```
In[10]:=  surfaceArea = 2 Pi r^2 + 2 Pi r h
Out[10]=                    2
          2 h Pi r + 2 Pi r
```

The first term on the right represents the area of the top and bottom. The second term represents the area of the side (which has the same area as a rectangle of length $2\pi r$ and height h).

The total length of the seam is

```
In[11]:=  seamLength = 4Pi r + h

Out[11]=  h + 4 Pi r
```

We will use centimeters as our units. The cost of the material is $costM = 0.01$ cents per square centimeter (100 cents divided by the 10000 square centimeters in a square meter). The seaming cost is $costS = 0.2$ cents per centimeter. The cost of the material is a surfaceArea and the cost of the seams is costS seamLength. So the total cost of the can is

```
In[12]:=  totalCost = costM surfaceArea + costS seamLength;
          costM = 0.01;
          costS = 0.2;
```

This cost expression has both r and h as variables. We eliminate the variable h by using the fact that the volume of the box is $1000 = \pi r^2 h$ (volume = base × height). Entering this fact into *Mathematica* yields a cost expression in terms of the single variable r.

```
In[15]:=  h = 1000/(Pi r^2);
          totalCost

Out[15]=      1000                         2000          2
          0.2 (----- + 4 Pi r) + 0.01 (---- + 2 Pi r )
                   2                        r
              Pi r
```

This cost expression is to be minimized over the interval $r > 0$. Plotting it, we observe that a minimum occurs somewhere between $r = 3$ and $r = 6$.

```
In[16]:=  Plot[c, {r, 1, 10}];
```

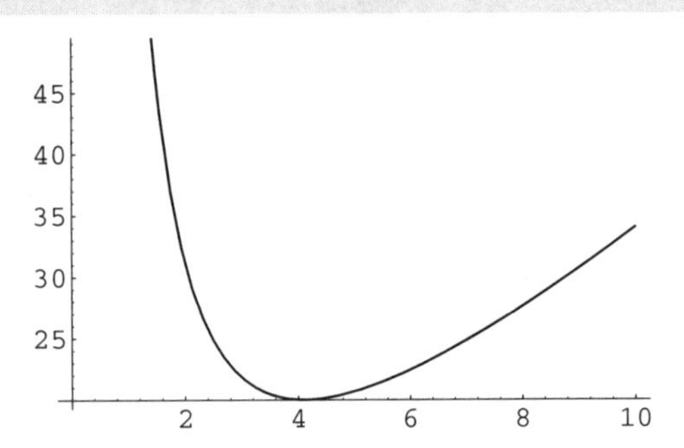

To minimize the cost expression, set its derivative Dc equal to zero, and solve for r. Assign the name rBest to the solution.

```
In[17]:=  DtotalCost = D[totalCost, r];
          rBest = FindRoot[DtotalCost = 0, {r, 3}]

Out[18]=  {r -> 4.10422}
```

The answer is the value for the radius (in centimeters) that minimizes the cost of a can whose volume is 1000 cubic centimeters. Substituting this value into the expression for h yields

```
In[19]:= N[h /. rBest]

Out[19]= 18.8969
```

Thus, the dimensions of the most economical can are 4.10422 centimeters by 18.8969 centimeters.

6.3 Summary

- You can *estimate* the absolute maximum or minimum of a *Mathematica* expression, `expr`, on a closed interval $a \leq x \leq b$ by plotting the expression and visually inspecting the plot.

```
Plot[expr, {x, a, b}];        (* Plot expr. *)
```

- You can *calculate* the absolute maximum or minimum of a *Mathematica* expression, `expr`, on a closed interval $a \leq x \leq b$ by using the functions `ConstrainedMax` or `ConstrainedMin`.

- You can use `FindMinimum` with an initial guess to find a local minimum of an expression.

- You can use `FindMaximum` with an initial guess to find a local maximum of an expression.

- You can use `D` in conjunction with `FindRoot` with an initial guess, to find a local maximum or local minimum. You must compute the derivative of the expression, set it equal to zero, and then solve for the roots of this equation:

```
Df = D[expr, x] (* Compute the derivative expression. *)
criticalPt = FindRoot(Df == 0, {x, c}]
             (* Locate the critical point.        *)
expr /. criticalPt
             (* Value of expr at the critical point. *)
```

  You should then determine the sign of the second derivative at the critical point to verify that it is a maximum or minimum as you suspect.

- If you are using the Notebook Front End version of *Mathematica*, you can see in the lower edge of a window the approximate coordinates of a point by clicking the mouse button on a graph, then holding down the ⟨COMMAND⟩ key and positioning the pointer above the point on the graph. This technique is useful for obtaining the initial guess as to the coordinates of a maximum or minimum that is needed for the functions `FindMinimum`, `FindMaximum`, and `FindRoot`.

- You can explicitly limit the range of values searched in the functions `FindMinimum`, `FindMaximum`, and `FindRoot` by giving two extra arguments, `xMin` and `xMax`. The search is terminated if the `x`-values being explored fall outside the range (`xMin, xMax`):

```
FindMinimum[expr, {x, xStart, xMin, xMax}]
FindMaximum[expr, {x, xStart, xMin, xMax}]
   FindRoot[expr, {x, xStart, xMin, xMax}]
```

- You can compare extrema in the interior of an interval with values at the endpoints a and b to select absolute extrema.

```
expr /. x -> a    (* Value of expr at left endpoint.  *)
expr /. x -> b    (* Value of expr at right endpoint. *)
```

6.4 Exercises

1. A metal box with a square base and no top holds 1000 cubic centimeters. It is formed by folding up the sides of the flattened pattern pictured here and seaming up the four sides. The material for the box costs \$1.00 per square meter, and the cost to seam the sides is 5 cents per meter. Find the dimensions of the box that costs the least to produce.

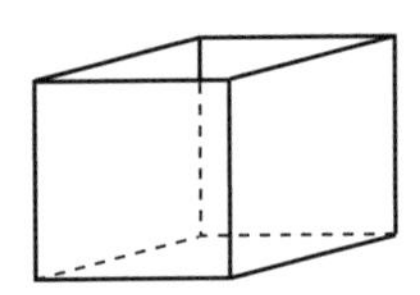

2. Find the point on the graph of $y = x^2 \tan(x)$ that is closest to the point $(2.1, 0.8)$. *Hint:* Use the distance formula to obtain a function that describes the distance between the point $(2.1, 0.8)$ and a typical point $(x, x^2 \tan(x))$ on the graph of y. Then, find its minimum using *Mathematica*.

3. Find the point on the parabola $y = \dfrac{x^2}{4} - 1$ that is closest to the point $(2, -1)$.

4. A rectangular movie theater is 100 feet long (from the front screen to the back row). The top and bottom of its screen are 40 feet and 15 feet from the floor, respectively. Find the position in the theater that has the largest viewing angle. *Hint:* Arrange a coordinate system with the origin at the floor directly under the screen. You are asked to find the position x where the angle $B - A$ is the largest (see figure). Instead of maximizing $B - A$, it is easier to maximize $\tan(B - A)$; this will lead to the same optimum value of x, since $\tan(\alpha)$ is an increasing function on $-\pi/2 < \alpha < \pi/2$. The subtraction formula for tangent will be needed.

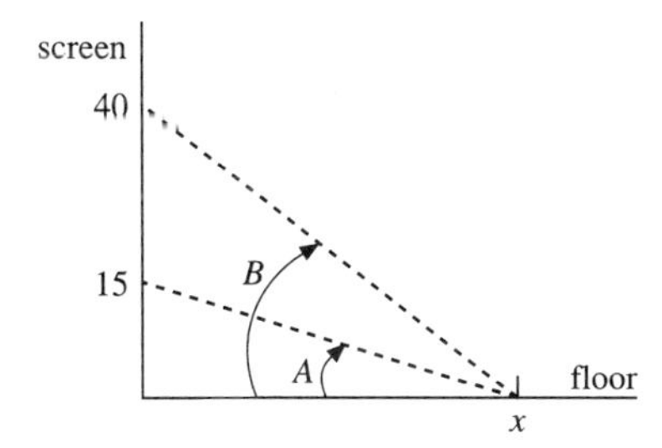

5. Repeat exercise 4, but this time assume that the floor of the theater has a gentle parabolic slope, given by the parabola $y = 0.001x^2$ for $0 \leq x \leq 100$ (the origin is located on the floor directly under the screen).

6. Repeat exercise 5, but assume that the floor of the theater has a steeper parabolic slope, given by the formula $y = 0.004x^2$.

7. A pipeline is to be constructed to connect a station on the shore of a straight section of coastline to a drilling rig that lies 5 kilometers down the coast and 2 kilometers out at sea. Find the minimum cost to construct the pipeline, given that the pipeline costs \$4,000,000 per kilometer to lay under water, and \$2,000,000 per kilometer to lay along shore.

8. Let v_1 be the velocity of light in air and v_2 the velocity of light in water. We know from physics that a ray of light travels from a point A in the air to a point B in the water via a path ACB that minimizes the time taken. The point of this problem is to derive Snell's law

$$\frac{\sin(\theta_1)}{\sin(\theta_2)} = \frac{v_1}{v_2}$$

where θ_1 (the angle of incidence) and θ_2 (the angle of refraction) are shown in the figure. This equation can then be used to calculate θ_2, provided θ_1, v_1, and v_2 are known. *Remark:* Use *Mathematica* to compute a derivative, if you wish; the rest of the solution is easier to do by hand.

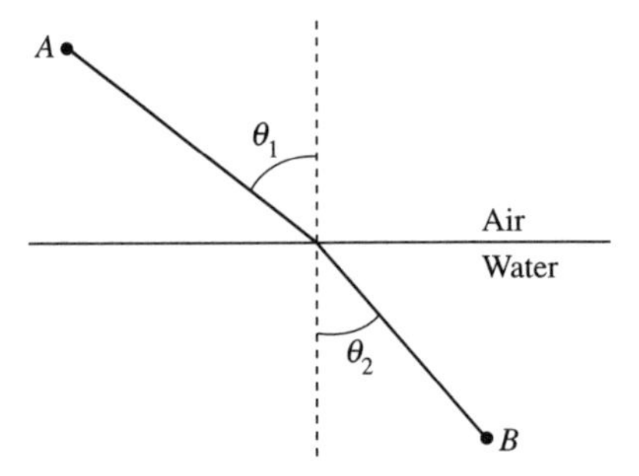

9. Now suppose the medium in exercise 8 has a curved surface (such as the surface of a glass lens). More specifically, suppose that a beam of light follows the line with the equation $y = -3x + 1$ and strikes a piece of glass whose outside boundary is given by the equation $y = -x^2$. Find the equation of the line that represents the refracted light beam as it travels through the glass. Assume that v_1/v_2 is 1.52.

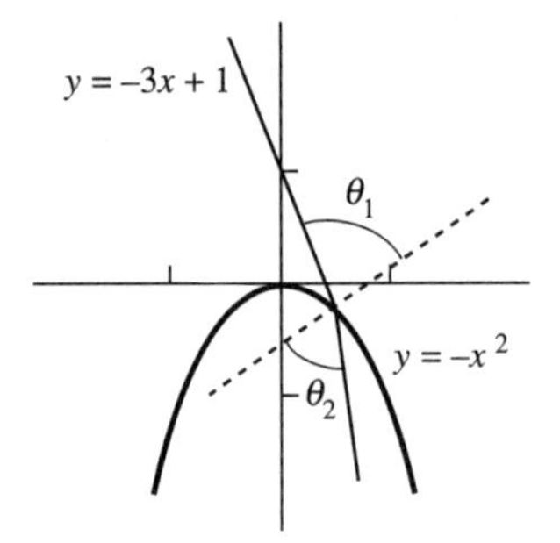

Hint: First, use *Mathematica* to find the point of intersection between the light beam and the parabolic surface of the glass. At this point, compute the angle between the normal to the parabolic surface and the incoming light beam. This angle is analogous to the angle θ_1 in exercise 8. You can now use Snell's law to calculate θ_2, and you can then determine the slope of the refracted light.

10. Suppose that a glass lens is formed by the intersection of the interiors of the following two ellipses: $(x-4)^2 + y^2/2 = 25$ and $(x+4)^2 + y^2/2 = 25$. A light beam strikes this lens from the right along the horizontal line $y = 2$. Find the equations of the line segments that describe the trajectories of the refracted light beam as it passes through the lens and as it passes through the air on the left side of the lens. Find the location where the refracted light beam hits the negative x-axis.

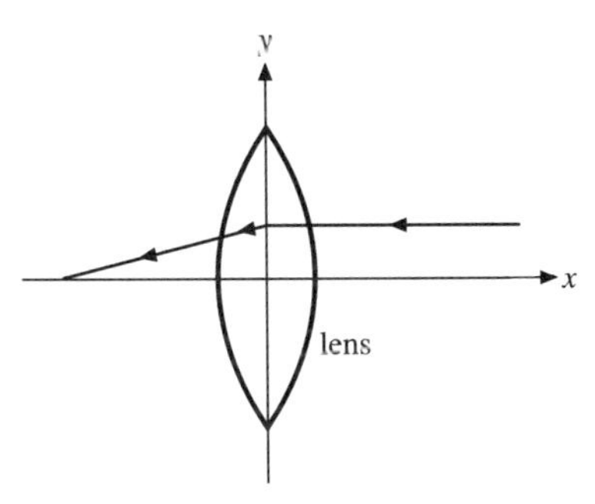

11. Two hallways of width a and b intersect at right angles. What is the length of the longest rigid rod that can be pushed on the floor around the corner of the intersection of these two halls?

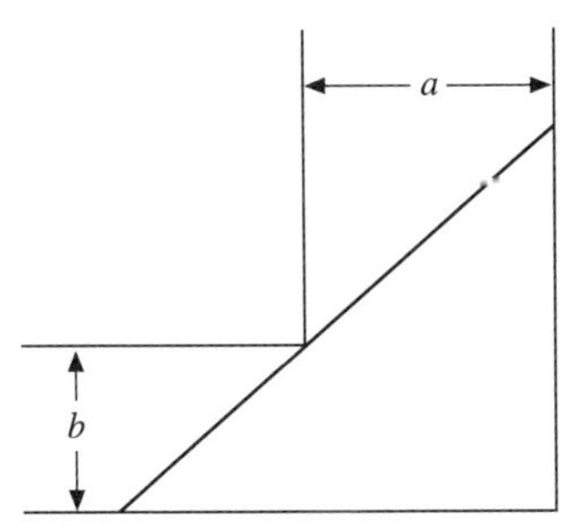

7 The Definite Integral

The integral may be used to calculate the area under a curve. The general idea is to divide up the area under the curve into the sum of areas of rectangles (called a *Riemann sum*). In the first section, we write *Mathematica* functions called `LeftBox` and `RightBox` to help visualize this process. In the second section, we compute some Riemann sums with the help of *Mathematica*'s `Sum` command. We also introduce the command `Integrate`.

7.1 Visualization of Riemann Sums

Consider the region under the graph of the function *f(x)*=x^2 over the interval $1 \leq x \leq 3$. First, divide this interval into n subintervals of equal length. Here, n is typically a large number such as 10 or 100 (later, n will approach ∞). The area under the graph of $f(x) = x^2$ is approximated by the sum of the areas of n rectangles where the base of a typical rectangle is one of the n subintervals and the height is the value of the function $f(x)$ at the left or right endpoint of the subinterval. To get a picture of $n = 10$ rectangles, we need to define a *Mathematica* function.

```
In[1]:=  LeftBox[fn_, {x_, a_, b_}, nRectangles_]:=
             Show[
                 Plot[fn, {x,a,b},
                     DisplayFunction->Identity],
                 LeftSidesAndTops[fn,{x,a,b}, nRectangles],
                 DisplayFunction->$DisplayFunction
             ]

         LeftSidesAndTops[fn_, {x_, a_, b_}, n_]:=
             Module[{dx, leftSides},
                 dx = N[(b-a)/n];
                 leftSides = Table[LeftSideAndTop[fn,a,dx,i],
                                         {i,0,n-1}];
             Join[leftSides,
                 {Graphics[{Line[{{b,0},{b,fn /.
                                             x->(b-dx)}}]}]}]
             ]
```

```
In[3]:=  LeftSideAndTop[fn_, a_, dx_, i_]:=
           Module[{yBtmLeft,yTopLeft,yTopRight,yBtmRight},
               yBtmLeft   = {a + i dx, 0};
               yTopLeft   = {a + i dx, fn /.
                                         x->(a + i dx)};
               yTopRight = {a + (i+1) dx, fn /.
                                         x->(a + i dx)};
               yBtmRight = {a + (i+1) dx, 0};
               Graphics[{Line[{yBtmLeft, yTopLeft}],
                         Line[{yTopLeft, yTopRight}],
                         Line[{yBtmRight, yTopRight}]
               }]
           ]
```

This function depicts an approximation of the area under the curve by a series of rectangles whose upper-left corners touch the curve. To see what this function does, type

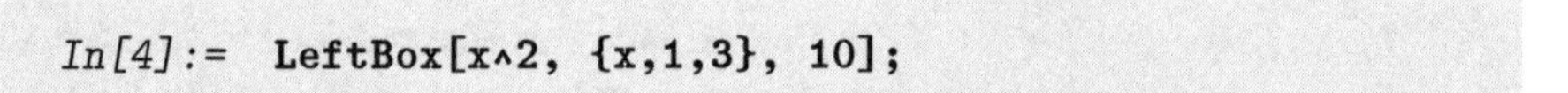

```
In[4]:=  LeftBox[x^2, {x,1,3}, 10];
```

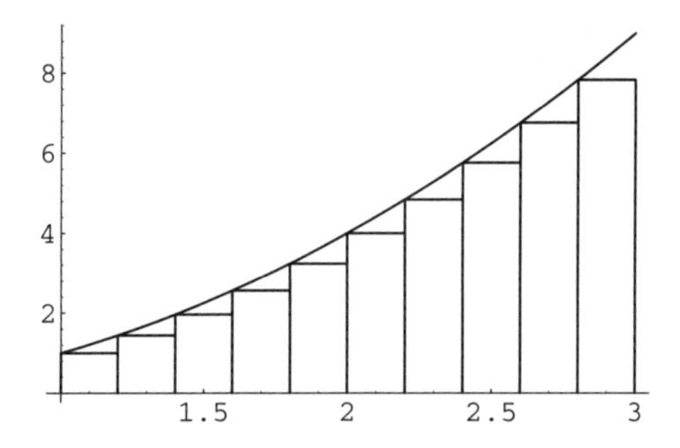

The `LeftBox` command draws the graph of $f(x)$ over the interval $1 \leq x \leq 3$, together with 10 rectangles; the height of each rectangle is the value of the function at the left point of the base of the rectangle. Since the graph of f is increasing over the interval $1 \leq x \leq 3$, the `leftbox` command gives rectangles whose areas under-approximate the area under the graph of f as the picture shows. To obtain rectangles whose heights are determined by the function values at the right endpoints, we define the `RightBox` command.

```
In[5]:=  RightBox[fn_, {x_, a_, b_}, nRectangles_] :=
           Show[
              Plot[fn, {x,a,b},
                 DisplayFunction -> Identity],
              RightSidesAndTops[fn,{x,a,b},nRectangles],
                 DisplayFunction -> $DisplayFunction
           ]

         RightSidesAndTops[fn_, {x_, a_, b_}, n_] :=
           Module[{dx, rightSides},
              dx = N[(b - a)/n];
              rightSides =
                 Table[RightSideAndTop[fn,a,dx,i],
                    {i,0,n-1}];
              Join[rightSides,
                 {Graphics[{Line[{{b,0},{b, fn /.
                                           x->b}}]}]}]
              ]

         RightSideAndTop[fn_, a_, dx_, i_]:=
           Module[{yBtmLeft,yTopLeft,yTopRight,yBtmRight},
              yBtmLeft  = {a + i dx, 0};
              yTopLeft  = {a + i dx, fn /.
                                 x->(a + (i+1) dx)};
              yTopRight = {a + (i+1) dx, fn /.
                                 x->(a + (i+1) dx)};
              yBtmRight = {a + (i+1) dx, 0};
              Graphics[{Line[{yBtmLeft, yTopLeft}],
                        Line[{yTopLeft, yTopRight}],
                        Line[{yBtmRight, yTopRight}]
              }]
           ]

In[8]:=  RightBox[x^2, {x, 1, 3}, 10];
```

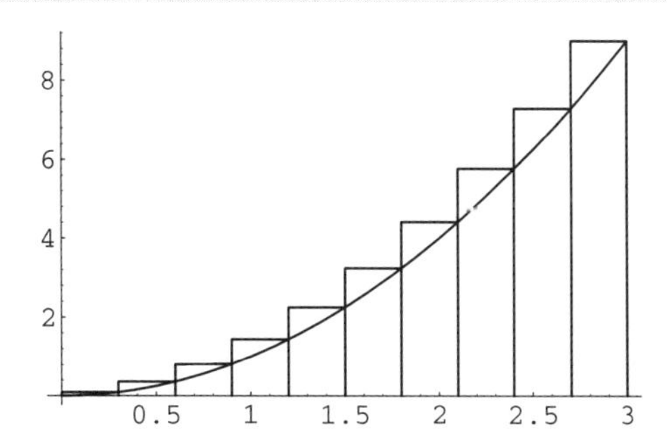

In this case, the `RightBox` command illustrates rectangles that over-estimate the area under the graph of f, since f is an increasing function on the interval $1 \leq x \leq 3$.

The value $n = 10$ can be changed to any positive integer. Try the `LeftBox` and `RightBox` commands with $n = 20, 50$, and 100. Note that, as the value of n gets larger, the sum of the areas of the rectangles more closely approximates the area under the graph of f. As n gets larger, the sum of the areas of the `LeftBox` rectangles approach the area under the graph from below, and the sum of the areas of the `RightBox` rectangles approach the area under the graph from above.

This same rectangle construction can be applied to any interval $a \leq x \leq b$ (rather than just the interval $1 \leq x \leq 3$). Try different values of a and b (and n). Note that, for negative values of a and b, f is a decreasing function on $a \leq x \leq b$, and therefore `LeftBox` rectangles will over-approximate and `RightBox` rectangles will under-approximate the area under the graph of f on such an interval.

For a nonnegative function, such as $f(x) = x^2$, this limit of the sum of the areas of rectangles (as the number of rectangles tends to infinity) is the definition of the integral of f from $x = a$ to $x = b$, and is denoted by the symbol

$$\int_a^b f(x)\,dx$$

7.2 Computation of the Definite Integral

We continue with the preceding example of the integral of $f(x) = x^2$ from $x = a$ to $x = b$.

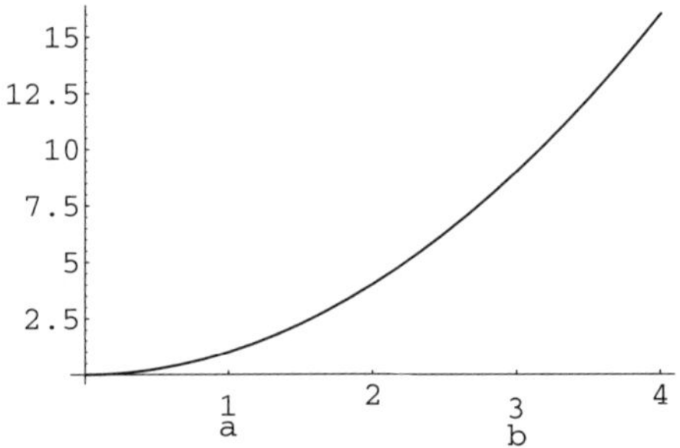

As in Section 7.1, the interval $a \leq x \leq b$ is divided into n subintervals, each of length

$$dx = \frac{b - a}{n}$$

Let x_i denote the right endpoint of the ith subinterval, where i is a counting index that runs from 1 to n. The area of the ith box is the product of the interval length dx and the height of the box $f(x_i)$. So the sum of the areas of the n-`RightBox` rectangles is

$$\sum_{i=1}^{n} f(x_i)\,dx$$

Before using *Mathematica* to compute this sum, we must derive a formula for x_i. Note the following pattern

$$x_1 = a + dx, \quad x_2 = a + 2\,dx, \quad x_3 = a + 3\,dx$$

and so forth. So, in general,

$$x_i = a + i\,dx$$

Now we can enter these formulae into *Mathematica*.

```
In[9]:=  Clear[f]
         f[x_] := x^2
```

Then, enter dx

```
In[11]:=  dx = (b-a)/n
Out[11]=        b - a
          dx = -----
                  n
```

The *Mathematica* command `Sum` computes the sum of the expression *expr* with the index variable i starting with $i_{\min}$ and going to $i_{\max}$.

```
Sum[expr, {i, imin, imax}]
```

We enter the above sum with the `Sum` command.

```
In[12]:=  Sum[f[a+i dx] dx, {i, 1, n}]
```

We can define a simple function that sums the areas of the rectangles produced by either `LeftBox` or `RightBox`.

```
In[13]:=  AreaOfLeftBoxes[fn_, {x_, a_, b_}, n_] :=
              Module[{dx = (b-a)/n},
                  Expand[Sum[(fn /. x->(a + i dx)) dx,
                              {i, 0, n-1}]]
              ];
In[14]:=  AreaOfRightBoxes[fn_, {x_, a_, b_}, n_]:=
              Module[{dx = (b-a)/n},
                  Expand[Sum[(fn /. x->(a + i dx)) dx,
                              {i, 1, n}]]
              ];
```

The `Sum` command is designed for evaluating sums over definite ranges. It cannot evaluate symbolic sums, i.e., sums where the range is specified in terms of a variable, such as the sum

$$\sum_{i=1}^{n} i^2$$

unless you load the package `SymbolicSum.m`. After loading this package with the command `Needs["Algebra`SymbolicSum`"]`, we can compute the closed form of the sums of the areas of the rectangles.

```
In[15]:= Needs["Algebra`SymbolicSum`"]
         area = AreaOfRightBoxes[x^2, {x, a, b}, n]

Out[16]=   3     3      3       2         2        3      3       2
         -a     b      a       a  b     a b       b      a       a  b
         --- + -- - ---- + ---- - ---- + ---- + --- - ---- -
          3     3      2       2        2       2     2 n     2 n
                    6 n     2 n      2 n     6 n

              2     3
           a b     b
           ---- + ---
           2 n    2 n
```

The *Mathematica* output represents the sum of the areas of n rectangles that approximates the area under the graph of $f(x) = x^2$ from $x = a$ to $x = b$. To get the precise area under the graph, take the limit of this expression as n approaches infinity.

```
In[17]:= Limit[area, n -> Infinity]

Out[17]=   3     3
         -a     b
         --- + --
          3     3
```

Note that this expression is the same as

$$F(b) - F(a)$$

where $F(x) = \frac{x^3}{3}$ is an antiderivative of $f(x) = x^2$. This result is an illustration of the *Fundamental Theorem of Calculus*, which states that the definite integral $\int_a^b f(x)\,dx$ (defined in terms of the sum of areas of rectangles) is the same as $F(b) - F(a)$, where F is an antiderivative of f.

As a further illustration of the Fundamental Theorem, try repeating the above procedure with the functions $f(x) = x^3$ and $f(x) = x^4$. The only *Mathematica* command that you must change is the one that involves the definition of the function f. The other statements can be reexecuted without change.

Mathematica has a command that integrates expressions. To integrate a function such as $f(x) = x^2$, enter

```
In[18]:=  Clear[f]
          f[x_] := x^2
          Integrate[f[x], {x, a, b}]

Out[20]=    3     3
          -a     b
          --- + --
           3     3
```

The antiderivative of f (or indefinite integral) can also be evaluated.

```
In[21]:=  Integrate[f[x], x]

Out[21]=   3
          x
          --
          3
```

Note that *Mathematica* does not insert the constant of integration.

Mathematica cannot find an antiderivative for every expression. For example, try integrating $\sqrt{x^5+1}$ with *Mathematica*.You have to approximate some definite integrals by adding up rectangles or trapezoids or via some other more sophisticated technique. *Mathematica* makes it easy to evaluate an approximation to a definite integral by using `NIntegrate`. For example, you can evaluate the integral $\int_1^2 \sqrt{x^5+1}\,dx$ by entering

```
In[22]:=  NIntegrate[Sqrt[x^5+1], {x, 1, 2}]

Out[22]=  3.1471
```

7.3 Summary

- The commands `LeftBox` and `RightBox`, defined in this chapter, allow you to visualize approximations to definite integrals. You must load these functions into your *Mathematica* session before you can use them.
- You can use *Mathematica* to set up a Riemann sum over the interval from a to b using n rectangles. To set up a right-sum, do the following:

```
Clear[f]
f[x_] := expr         (* Define f to be a function in x. *)
dx = (b-a)/n;         (* Compute width of the rectangle. *)
area = Sum[f[a + i dx] dx, {i, 1, n}]
                      (* Attempt to sum in closed form.  *)
```

If, initially, *Mathematica* cannot evaluate the sum in closed form, try loading the package `Needs["Algebra`SymbolicSum`"]` and re-computing the previous steps.

- You can use the symbolic closed form of the Riemann sum to compute the area under the curve, by taking the limit as the number of boxes goes to infinity.

```
Limit[area, n -> Infinity]
```

- You can set up Riemann sums using the functions `AreaOfLeftBoxes` and `AreaOfRightBoxes` that we defined in Section 7.2.
- You can use `Integrate` to compute both definite and indefinite integrals. The syntax is as follows:

```
Integrate[f[x], x]
Integrate[f[x], {x, a, b}]
```

If *Mathematica* cannot compute a definite integral whose limits are numeric, you may be able to obtain an approximate answer using `NIntegrate`.

```
NIntegrate[f[x], {x, a, b}]
```

- Recall that, when *Mathematica* computes an indefinite integral, it does not add a constant of integration.

7.4 Exercises

1. Type in the definition of `LeftBox` given in Section 7.1.

   (a) Use the `LeftBox` command with 20 rectangles to display the Riemann sum approximation to

$$\int_0^2 \sqrt{8 - x^3}\, dx$$

Does this set of rectangles over-approximate or under-approximate the integral?

(b) Set up and evaluate the Riemann sum represented by the `LeftBox` command for $n = 20, 40$, and 80 rectangles. (*Hint:* Use `AreaOfLeftBoxes`).

(c) Compare the values of the Riemann sums in part (b) with the value that you obtained by using *Mathematica*'s `NIntegrate` command. As the number of rectangles doubles, what happens to the difference between the value of the Riemann sum and the value of the integral?

2. Repeat exercise 1 using the `RightBox` command (as well as setting up and evaluating the corresponding Riemann sums for $n = 20, 40$, and 80 rectangles).

3. Repeat exercises 1 and 2 with the function $f(x) = \dfrac{x^2 - 1}{x^4 + 1}$.

4. Use *Mathematica* to evaluate the following integrals.

(a) $\displaystyle\int_1^3 x\sqrt{x^2+1}\,dx$

(b) $\displaystyle\int \cos^4(x)\,dx$

(c) $\displaystyle\int \sin^6(x)\,dx$

5. There are two ways to evaluate the integral $\int \sec^2(x)\tan(x)\,dx$ by u substitution: (1) by letting $u = \sec(x)$ and (2) by letting $u = \tan(x)$. Based on *Mathematica*'s answer to this integral, which one is *Mathematica* using? These two methods of substitution lead to apparently different answers. Are they, in fact, the same? If they are different, then how do you explain the fact that they are answers to the same problem?

6. Express each of the following limits as an integral. Then, use *Mathematica* to evaluate it. (The first two integrals are provided for you.)

(a) $\displaystyle\lim_{n\to\infty}\sum_{i=1}^{n}\tfrac{1}{n}\sqrt{1+\tfrac{i^2}{n^2}} = \int_0^1 \sqrt{1+x^2}\,dx$

(b) $\displaystyle\lim_{n\to\infty}\sum_{i=1}^{n}\tfrac{2}{n}\sqrt{1+\tfrac{4i^2}{n^2}} = \int_0^2 \sqrt{1+x^2}\,dx$

(c) $\displaystyle\lim_{n\to\infty}\sum_{i=1}^{n}\tfrac{\pi}{n}\sin^2\left(\tfrac{i\pi}{n}\right)$

(d) $\displaystyle\lim_{n\to\infty}\sum_{i=1}^{n}\frac{2}{n}\left(\left(1+\frac{2i}{n}\right)^3+3\left(1+\frac{2i}{n}\right)\right)$

7. *Area of Texas.* From exercise 25 in Chapter 2, the northern and southern boundaries of the state of Texas are given by the following data.

```
north = {{0,0},{3,0},{3,4.5},{6,4.5},{6,2.2},{7,2.1},
         {8,1.8},{9,1.9},{10,1.8},{11,1.7},{11,-2.2}};
south = {{0,0},{1,-1.1},{2,-2.5},{3,-2.9},{4,-2.3},
         {5,-2.8},{6,-4.4},{7,-5.8},{8,-6.1},{9,-3.3},
         {10,-2.8},{11,-2.2}};
```

(Enter the data into *Mathematica.*) Here, the origin is the western corner of Texas (near El Paso), and the x-axis is the extension of the east-west border between New Mexico and Texas. Each unit represents approximately 69 miles.

Use these data to approximate the area of Texas by using a Riemann sum formed from rectangles whose widths are 1 unit and whose heights are determined by the second coordinates of the data. *Hint:* To refer to the entries in a list, use brackets [[]]. For example, `south[[3]]` refers to the point $[2, -2.5]$, and `south[[3,2]]` refers to the second entry of this point, -2.5. So to (symbolically) sum all second entries of the points on this list, enter `Sum[south[[i, 2]], {i, 1, 11}]`.

8 Area and Volume

The first section of this chapter uses *Mathematica* to compute areas. The second section discusses how to compute volumes.

8.1 Area

We present two examples. In the first example, we compute the area between a curve and the x-axis, and in the second example, we compute the area between the two curves.

Example 8.1 Find the area that lies between the curve given by the following function and the x-axis.

```
In[1]:=  f[x_] := -0.128 x^3 + 1.728 x^2 - 5.376 x + 2.864
```

Solution. First, take a look at the graph of f to see where it crosses the x-axis. In this case, we need only check the interval $-2 \leq x \leq 10$.

```
In[2]:=  Plot[f[x], {x, -2, 10}];
```

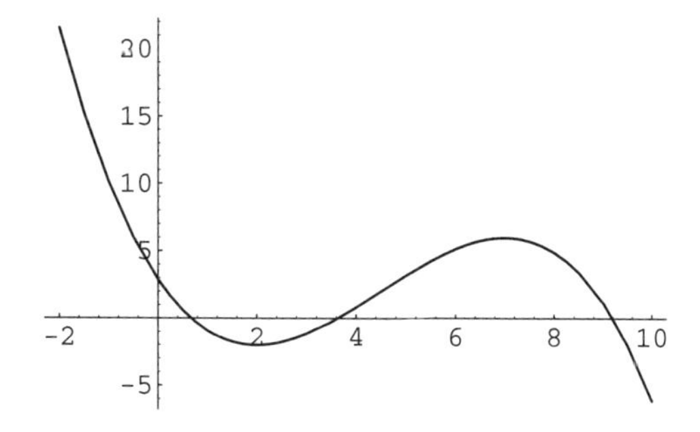

The graph of f crosses the x-axis at three points—near 0, between 3 and 4, and between 9 and 10. Use `NSolve` to find these roots. Assign the name `roots` to the roots, so that we can easily refer to them later on.

```
In[3]:=  roots = NSolve[f[x] == 0, x]

Out[3]=  {{x -> 0.669778}, {x -> 3.63176}, {x -> 9.19846}}
```

We can find all the roots by using `NSolve` because f is a polynomial. The variable `roots` is set equal to a list. We can refer to each of its elements by typing `x /. roots[[1]]`, `x /. roots[[2]]`, and so on. From the plot, it is evident that the graph of f is below the x-axis between the first root and the second root, and above the x-axis between the second root and the third root. Therefore, to evaluate the area between the curve and the x-axis, we compute the following integral:

$$-\int_{\text{root1}}^{\text{root2}} f\,dx + \int_{\text{root2}}^{\text{root3}} f\,dx$$

In *Mathematica*, we evaluate this integral by entering the commands

```
In[4]:=  root1 = x /. roots[[1]];
         root2 = x /. roots[[2]];
         root3 = x /. roots[[3]];
         -Integrate[f[x], {x, root1, root2}] +
             Integrate[f[x], {x, root2, root3}]

Out[7]=  25.0502
```

The area of interest is about 25.05 square units.

Example 8.2 Compute the area between the graph of f (defined in Example 8.1) and the graph of g, where

```
In[8]:=  g[x_] := 0.08 x^3 - 0.84 x^2 + 1.44 x + 4.32
```

Solution. First, plot f and g on the same coordinate axes with the command

```
In[9]:=  Plot[{f[x], g[x]}, {x, -2, 10}];
```

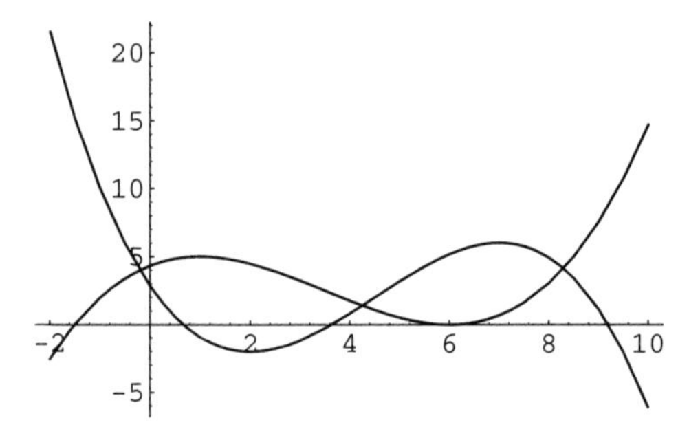

The plot shows three points of intersection. Use `NSolve` to find the x-coordinate of these points.

```
In[10]:=  roots = NSolve[f[x] == g[x], x]

Out[10]=  {{x -> -0.198527}, {x -> 4.2518}, {x -> 8.29288}}
```

The graph of f is below the graph of g between the first and second points of intersection (which we will call intPt1 and intPt2, respectively), and the graph of g is below the graph of f between the second and third points of intersection. Therefore, we compute the area between the graphs of f and g by using the following integral.

$$\int_{\text{intPt1}}^{\text{intPt2}} (g - f)\, dx + \int_{\text{intPt2}}^{\text{intPt3}} (f - g) dx$$

In *Mathematica*, this integral is entered as

```
In[11]:=  intPt1 = x /. roots[[1]];
          intPt2 = x /. roots[[2]];
          intPt3 = x /. roots[[3]];
          Integrate[g[x] - f[x], {x, intPt1, intPt2}] +
          Integrate[f[x] - g[x], {x, intPt2, intPt3}]

Out[15]=  33.9503
```

8.2 Volume

If we wish to calculate a volume by slicing, the first step is to define a function that represents the cross-sectional area. Then, we integrate this function over the appropriate interval.

Example 8.3 Consider the region that is bounded above by the curve $y = -x^2 + 5x - 2$ and below by the line $y = x$. Find the volume of the solid that we obtain by revolving this region about the x-axis.
Solution. First, define the functions $f = -x^2 + 5x - 2$ and $g = x$ in *Mathematica*, and graph these expressions with a plot command.

```
In[16]:=  Clear[f, g]
          f[x_] := -x^2 + 5x - 2
          g[x_] := x
          Plot[{f[x], g[x]}, {x, 0, 4}];
```

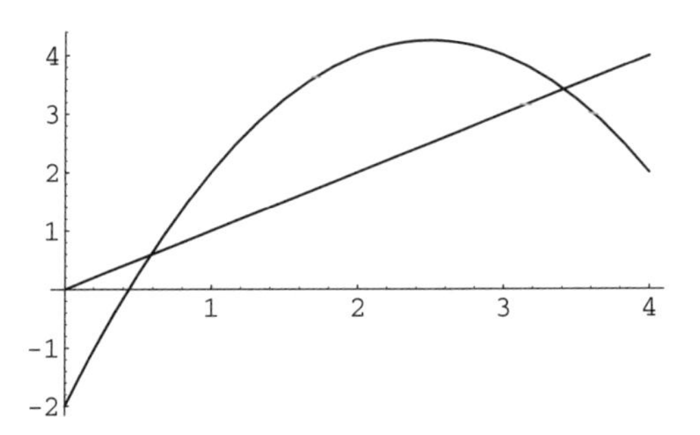

From the plot, it is evident that f and g cross at two points: one with an x-coordinate between $x = 0$ and $x = 1$, and the other with an x-coordinate between $x = 3$ and $x = 4$. We can calculate these roots by using `FindRoot`. (*Note:* We could have used `NSolve` to find the roots, because f and g are both polynomial expressions.) Then, we assign them the names `a` and `b`.

```
In[20]:=  a = x /. FindRoot[f[x] == g[x], {x, 0}]

Out[20]=  0.585786

In[21]:=  b = x /. FindRoot[f[x] == g[x], {x, 3}]

Out[21]=  3.41421
```

The cross-section of this solid is in the shape of a *washer* whose outside radius is given by f and whose inside radius is given by g.

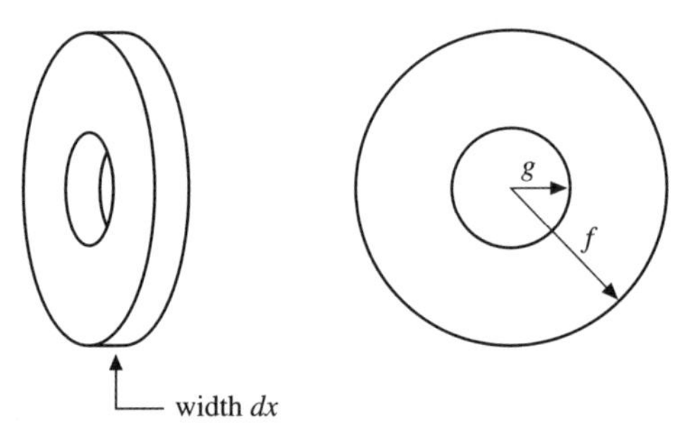

The area of the cross-section is given by $\pi(f(x)^2 - g(x)^2)$. We obtain the total volume of this solid by summing (integrating) this expression over the interval $a \leq x \leq b$. *Mathematica* can do this integral easily.

```
In[22]:=  Integrate[Pi (f[x]^2 - g[x]^2), {x, a, b}]

Out[22]=  21.1189 Pi
```

Now use `N` to compute an approximate value of this integral.

```
In[23]:=  N[%]

Out[23]=  66.3471
```

Example 8.4 Find the volume of the solid that we obtain by revolving this region about the y-axis.
Solution. In this case, finding the area of the cross-section (which is a washer perpendicular to the y-axis) is more difficult (try this problem as a challenge). An easier method is to calculate the volume by the method of cylindrical shells. The idea is to slice the volume by shells whose radius, height, and thickness are given by x, $f - g$, and dx, respectively.

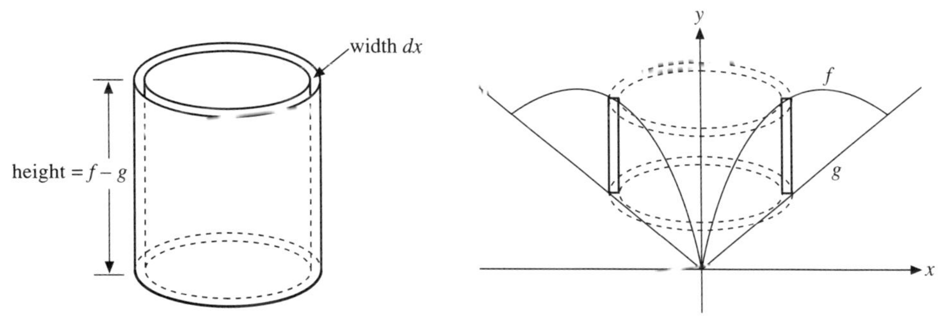

The volume of this shell is given by $2\pi x(f(x) - g(x))\,dx$. We obtain the total volume of this solid by summing (integrating) this expression over the interval $a \leq x \leq b$.

```
In[24]:=  Integrate[2Pi x (f[x] - g[x]), {x, a, b}] // N

Out[24]=  47.3908
```

8.3 Summary

- To find the area between a curve f and the x-axis,

  (1) Use `NSolve` or `FindRoot` (if f is not a polynomial) to find the roots of f, i.e., the points where the curve crosses the x-axis.

  (2) Use `Integrate` on the difference f between successive roots.

  (3) Add the values returned by `Integrate` that correspond to regions that lie above the x-axis and subtract the values returned by `Integrate` that correspond to regions that lie below the x-axis.

- To find the area between two curves f and g,

  (1) Use NSolve or FindRoot (if f or g is not a polynomial) to find their points of intersection.

  (2) Use Integrate on the difference $f - g$ between successive points of intesection.

  (3) Add the absolute values of the values returned by Integrate, i.e.,

```
Abs[Integrate[f[x] - g[x], {x, cross1, cross2}]] +
Abs[Integrate[f[x] - g[x], {x, cross2, cross3}]] + ...
```

- You can use Integrate to compute volumes of solids of revolution by slicing solids into annular washers or cylindrical shells.

8.4 Exercises

1. Find the area of the region that is bounded above by the curves $y = 7\ln(x)$ and $y = 4 - x^3 - x$ and below by the x-axis.

2. Consider the region defined in exercise 1.

   (a) Find the volume of the solid that you obtain by revolving this region about the x-axis.

   (b) Find the volume of the solid that you obtain by revolving this region about the y-axis.

   (c) Find the volume of the solid that you obtain by revolving this region about the line $x = 4$.

   (d) Find the volume of the solid that you obtain by revolving this region about the line $y = 4$.

3. Consider the solid that you obtain by revolving the region bounded by the graph of $y = 16 - x^2$ and the coordinate axes about the x-axis. Use the Sum and N commands to approximate the volume of this solid by summing the volumes of 10, 20, 100, and 1000 inscribed shells of equal thickness. Now repeat with circumscribed shells.

4. Compute the volume of the solid in exercise 3 by integration. Compare your answers.

5. This problem is a continuation of exercise 10 in Chapter 5. In that exercise, the northern boundary of a plot of land is described by the following table (each unit represents 100 feet).

x-value	0.0	1.0	2.0	3.0	4.0
y-value	0.0	2.1	3.2	2.4	1.7

where the x-axis represents the (flat) southern boundary with the origin located at its western corner. Use these data to compute an approximation to the area of this plot of land using rectangles.

6. Do exercise 10 from Chapter 5, or load the answer from a file if you have already done that exercise. Use the quadratic and cubic functions described therein to evaluate the area of this plot of land.

7. *The volume of a JELL-O® mold.* Consider the region bounded below by the x-axis and above by the parabola that passes through the points $(8, 0)$, $(10, 4)$, and $(12, 0)$.

   (a) Find the volume of the solid that you obtain by revolving this region about the y-axis.

   (b) Find the volume of the solid that you obtain by revolving this region about the line $x = 2$.

8. In this exercise, you are to calculate the approximate volume of a bowl of depth 3 feet with circular cross-sections. The measurements of the radius of the cross-sections versus depth are given in the following table (in feet).

Depth	0.00	0.50	1.00	1.50	2.00	2.50	3.00
Radius	2.00	1.85	1.60	1.40	1.05	0.55	0.00

Find an approximate value of the volume of this bowl by adding the volumes of the shells of width 0.5 feet and radius given by the values in the table.

9. Find a third-degree polynomial whose graph contains every other data point in the table in exercise 8. Use this polynomial to compute an approximate volume for the bowl. Compare your answer with that of the previous exercise. *Hint:* Think of the x-axis as representing depth, and enter $p(x) = ax^3 + bx^2 + cx + d$ as a function in *Mathematica*. Then, solve the four equations $p(0) = 2$, $p(1) = 1.6$, $p(2) = 1.05$, and $p(3) = 0$ for the unknowns $a, b, c,$ and d.Graph this polynomial to see whether its graph contains these data points.

10. Suppose you form a circular doughnut by revolving a circle of radius r centered at $x = a$, $y = 0$ about the y-axis ($a > r$). Find the formula for the volume of this doughnut.

   (a) Find the equation of the circle described above.

   (b) Review the cylindrical-shell technique for finding volumes. Then, set up the integral for the volume of this doughnut.

   (c) Use *Mathematica*'s `Integrate` command to evaluate the integral from part (b).

   (d) Evaluate the volume in the special case where $a = 3$ and $r = 1$.

   (e) To see a three-dimensional picture of this doughnut, issue the following *Mathematica* commands.

```
x[s_, t_] := Cos[s] (3 + Cos[t])
y[s_, t_] := Sin[s] (3 + Cos[t])
z[t_] := Sin[t]
```

See whether you can figure out why these three equations parameterize the doughnut as the parameters s and t vary from 0 to 2π. Then, type

```
ParametricPlot3D[[x[s, t], y[s, t], z[t]],
    {s, 0, 2Pi}, {t, 0, 2Pi}];
```

9 Techniques of Integration

There are many techniques of integration. In this chapter, we will demonstrate how to use three techniques with *Mathematica*:

1. Integration by substitution (change of variables): $\int f(g(x))g'(x)\,dx = \int f(u)\,du$.
2. Integration by parts: $\int u(x)\,v'(x)\,dx = u(x)v(x) - \int v(x)\,u'(x)\,dx$.
3. Integration of rational functions by partial fractions.

You might wonder why we are bothering to discuss techniques of integration when *Mathematica* comes equipped with the built in function `Integrate`. It turns out that `Integrate` has its limitations. For example, try using *Mathematica* to integrate the function $f(x) = \sin(x^2)\ln(x)$.

```
Integrate[Sin[x^2] Log[x], x]
```

Mathematica is stumped. In such a situation, you will have to exert more control over how *Mathematica* approaches the integral. The above three techniques are often of assistance. We will show you how to implement these techniques in *Mathematica* and how to use them on real problems.

9.1 Integration by Substitution (Change of Variables)

Suppose that we want to compute the integral

$$A = \int \frac{1}{\sqrt{4+9x^2}}\,dx$$

by making a change of variables. Actually, *Mathematica* can do this calculation without changing variables, but it's easier to understand a technique when it is demonstrated on a simple example.

First, we identify a relation between the old variable x and a new variable. In this case, take

$$3x = 2\tan\theta$$

where the new variable is θ. Compute the differential of this relation, and solve for dx. We now have

$$\begin{aligned} 3\,dx &= 2\sec^2\theta\,d\theta \\ dx &= \frac{2}{3}\sec^2\theta\,d\theta \end{aligned}$$

Next, substitute these definitions into the integral.

$$A = \int \frac{2\sec^2\theta}{3\sqrt{4+4\tan^2\theta}}\,d\theta = \frac{1}{3}\int \sec\theta\,d\theta = \frac{1}{3}\ln|\sec\theta + \tan\theta| + C$$

Finally, substitute back to obtain

$$A = \frac{1}{3}\ln\left|\frac{\sqrt{4+9x^2}}{2} + \frac{3}{2}x\right| + C$$

To do this computation using *Mathematica*, we define a function for changing the variable of integration. Let us call it `ChangeVariable`. This function takes four arguments: the first argument is the integrand, the second is the original variable of integration, the third is the relation between the old and new variables, and the fourth is the name of the new variable.

```
In[1]:=  (* Definite Integration *)
         IntegrateBySubstitution[integrand_, {x_,a_,b_},
               fx_ == gu_, u_]:=
            Integrate[
               ChangeVariable[integrand, x ,fx == gu, u],
               ChangeLimits[{x, a, b}, fx == gu, u]
          ]

         (* Indefinite Integration *)
         IntegrateBySubstitution[integrand_,x_,fx_==gu_,u_]:=
            Integrate[
               ChangeVariable[integrand,x,fx==gu,u],u] //.
                     u -> Solve[fx == gu, u][[1,1,2]]

         ChangeVariable[xIntegrand_, x_, x_ == gu_, u_]:=
            ((xIntegrand //. x->gu) D[gu, u]) /;
                                                FreeQ[gu,x]

         ChangeVariable[integrand_, x_, gu_ == fx_, u_]:=
            ChangeVariable[integrand, x, fx == gu, u] /;
               FreeQ[gu, x] && FreeQ[fx, u]

         ChangeVariable[xIntegrand_, x_, fx_ == gu_, u_]:=
            Module[{solns},
               solns = Solve[fx == gu, x];
               Map[ChangeVariable[xIntegrand, x,
                  x==(x /. #), u]&, solns]
            ] /; FreeQ[gu,x] && FreeQ[fx,u]
```

```
In[6]:=  ChangeLimits[{x_, a_, b_}, fx_ == gu_, u_]:=
             Module[{inverseFns},
                 inverseFns = Solve[fx == gu, u];
                 If[Length[inverseFns] > 1,
                     Message[
                         ChangeLimits::inverseIsNot1To1],
                    {u, inverseFns[[1,1,2]] /. x -> a,
                     inverseFns[[1,1,2]] /. x -> b}
             ]
          ]
         ChangeLimits::inverseIsNot1To1 = "The limits are
             ambiguous because the inverse substitution is
             not 1 to 1.";
```

So, to compute $\int 1/\sqrt{4+9x^2}\,dx$, execute a change of variables (by using ChangeVariable), and evaluate the integral (using IntegrateBySubstitution).

```
In[8]:=  IntegrateBySubstitution[1/Sqrt[4 + 9 x^2], x,
             x == (2/3) Tan[t], t]

Solve::ifun: Warning: Inverse functions are being used by
     Solve, so some solutions may not be found.

Out[8]=          3 x
         ArcSinh[---]
                  2
         ------------
               3
```

This is the same result that we obtain by using Integrate directly.

```
In[9]:=  Integrate[1/Sqrt[4 + 9 x^2], x]

Out[9]=          3 x
         ArcSinh[---]
                  2
         ------------
               3
```

This is the answer without the additive constant. *Mathematica*'s answer is different from the one we obtained by hand. But recalling the formula $\sinh^{-1}(w) = \ln\left(\sqrt{w^2+1}+w\right)$, we see that they are equivalent. However, as with any indefinite integral, check the answer by differentiating the results. Thus, executing

```
In[10]:=  D[%, x]

Out[10]=          1
          ----------------
                         2
                      9 x
          2 Sqrt[1 + ----]
                       4

In[11]:=  Simplify[%]

Out[11]=         1
          --------------
                       2
          Sqrt[4 + 9 x ]
```

gives back the integrand of the original integral A.

If the integral becomes more complicated when you apply the `ChangeVariable` command, go back and try a different substitution or a different integration trick.

9.2 Integration by Parts

Suppose that we want to compute the integral $\int x \sin x \, dx$ by using integration by parts. First, identify $u(x)$ and $v'(x)$. In this case, take

$$u = x \text{ and } v'(x) = \sin x$$

Then compute $u'(x)$ and $v(x)$. We then have

$$u'(x) = 1 \text{ and } v(x) = -\cos x$$

The integration-by-parts formula $\int u(x)\, v'(x)\, dx = u(x)v(x) - \int v(x)\, u'(x)\, dx$ gives

$$\int x \sin x \, dx = -x \cos x + \int \cos x \, dx = -x \cos x + \sin x + C$$

To do this computation using *Mathematica*, we define a function that implements integration by parts. Let us call it the `IntegrateByParts` command and make it take three arguments: the first argument is the integrand, the second is the variable of integration or the range over which the integral is to be computed, and the third is the part of the integrand that will be taken as $u(x)$.

```
In[12]:=  IntegrateByParts[integrand_,{x_,a_,b_},uExpr_]:=
             Module[{u, v, uprime, vprime},
                 u = uExpr;
                 vprime = Simplify[integrand/u];
                 v = Integrate[vprime,x];
                 uprime = D[u, x];
                 (u v //. x -> b) - (u v //. x- > a) -
                     Integrate[v uprime, {x, a, b}]
             ]

          IntegrateByParts[integrand_, x_, uExpr_]:=
             Module[{u, v, uprime, vprime},
                 u = uExpr;
                 vprime = Simplify[integrand/u];
                 v = Integrate[vprime, x];
                 uprime = D[u, x];
                 u v - Integrate[v uprime, x]
             ]
```

So, to compute $\int x \sin x \, dx$, we load this definition into *Mathematica* and then execute:

```
In[14]:=  aParts = IntegrateByParts[x Sin[x], x, x]

Out[14]=  -(x Cos[x]) + Sin[x]
```

If the integral becomes more complicated when you apply the command `IntegrateByParts`, go back and try a different u or a different integration trick.

9.3 Integration of Rational Functions by Partial Fractions

You can often best tackle integrals of rational functions by first finding the partial fraction expansion of the function and then integrating each term in the resulting sum separately. In *Mathematica*, the function `Apart` returns the partial fractions. The command `Apart` takes two arguments: the first is the rational function to be expanded, and the second is the variable with respect to which the partial fraction expansion is to be performed. The second argument is not necessary if the function contains a single variable. To evaluate

$$\int \frac{x^2 - 3x + 1}{x^3 + x^2 - 2x} \, dx$$

define the integrand

```
In[15]:=  f[x_] := (x^2 - 3x + 1)/(x^3 + x^2 - 2x)
```

Find its partial fraction expansion

```
In[16]:=  Apart[f[x], x]

Out[16]=      -1          1         11
          ---------- - --- + ---------
          3 (-1 + x)   2 x   6 (2 + x)
```

Construct the integral, and find its value

```
In[17]:=  Integrate[%, x]

Out[17]=  -Log[-1 + x]   Log[x]    11 Log[2 + x]
          ------------ - ------ + -------------
               3           2            6
```

This procedure can be bundled up into a single function the way we did for integration by substitution and integration by parts:

```
IntegrateByPartialFractions[integral_, {x_, a_, b_}]:=
    Module[{partialfractions},
        partialfractions = Apart[integral, x];
        Integrate[partialfractions, {x, a, b}]
    ]

IntegrateByPartialFractions[integral_, x_]:=
    Module[{partialfractions},
        partialfractions = Apart[integral, x];
        Integrate[partialfractions, x]
    ]
```

9.4 Summary

At this point, you should be aware that you need not rely on just the built-in integration capabilities of *Mathematica*: you can define your own integration procedures. *Mathematica*'s functions for manipulating algebraic expressions are useful for this purpose. In this chapter we used `ReplaceAll`, `ReplaceRepeated`, and `Apart`. We have shown you how to write a general purpose `ChangeVariable` function, and how to rewrite the form of the integrand using `Apart`. Many other techniques for integration can be implemented in a similar fashion. Once you load in the definitions for `IntegrateBySubstitution`, `IntegrateByParts`, and `IntegrateByPartialFractions`, you will be able to guide *Mathematica* to do integrals that it cannot do on its own.

9.5 Exercises

1. Evaluate the integral

$$\int \frac{x^2 \tan^{-1}(x)}{(1+x^2)^2}\,dx$$

Then use the ChangeVariable, value, /., and Simplify commands. Be sure to check your answer.

2. Evaluate the integral $\int x \ln(x\sqrt{1+x^2})\,dx$. Then, use the IntegrateByParts command with $u = x$. Finally, use the IntegrateByParts command with $u = \ln(x\sqrt{1+x^2})$. Be sure to check your answer. Sometimes, you may have to apply integration by parts more than once to evaluate an integral successfully.

3. Use the IntegrateByParts command to compute $\int x^2 e^{-x}\,dx$ and $\int_1^3 x^2 e^{-x}\,dx$.

4. As a more sophisticated example of integration by parts, evaluate the integral

$$\int e^x \sin x\,dx$$

Start by giving a name to this integral and choosing $u = e^x$.

5. Find the partial fraction expansion for $f = \dfrac{x^8+2x-1}{(x-1)^3(x^2+3)^2}$. Then, use it to evaluate $\displaystyle\int \frac{x^8+2x-1}{(x-1)^3(x^2+3)^2}\,dx$.

Use *Mathematica* to compute the following integrals.

6. $\displaystyle\int \frac{x}{x^6+1}\,dx$

7. $\displaystyle\int \frac{\tan(\sqrt{x})}{2\sqrt{x}}\,dx$

8. $\displaystyle\int \sin^4(x)\cos^2(x)\,dx$

9. $\displaystyle\int_{-1}^{3} \frac{x^3+x}{\sqrt{1+x^2}}\,dx$

10. $\displaystyle\int_{-\pi/4}^{\pi/4} \sqrt{\tan^2(x)+1}\,dx$

11. $\int_{-3}^{-1} \sqrt{x^4 + x^2}\, dx$. We saw in Chapter 7 that *Mathematica* does not integrate this correctly. Try changing variables to help *Mathematica* evaluate the integral.

10 Sequences and Series

In this chapter, we show how to use *Mathematica*'s graphing capabilities to explore the behavior of sequences. We also show the commands `Table` and `Sum` to specify sequences and to sum the terms.

To construct a sequence of numbers whose individual terms are given by a formula, first enter the formula that describes the terms as a function of `n`.

```
In[1]:=  Clear[a, n]
         a[n_] := 1/n^2
```

To have *Mathematica* construct the first five terms of this sequence, use the `Table` command. Its first argument is an expression for the terms of the sequence. Its second argument tells *Mathematica* which terms to construct.

```
In[3]:=  Table[a[n], {n, 5, 9}]

Out[3]=   1    1    1    1    1
         {--,  --,  --,  --,  --}
          25   36   49   64   81
```

The preceding command instructs *Mathematica* to construct a list consisting of the values {`a[5]`, `a[6]`, `a[7]`, `a[8]`, `a[9]`}. Note that the index variable `n` assumes integer values from 5 to 9.

10.1 Limits of Sequences

Let's examine the limiting value of a sequence as $n \to \infty$. *Mathematica* is good at calculating the limiting value and displaying the sequence graphically. Let's examine the behavior of the sequence $b_n = (2n - 1)/(3n + 6)$ by plotting the terms of this sequence as a function of the parameter n.

```
In[4]:=  Clear[b, n]
         b[n_] := (2n - 1)/(3n + 6)
         terms = Table[{n, b[n]}, {n, 1, 50}];
```

These *Mathematica* commands construct the sequence of lists that are the Cartesian coordinates of the points (n, b_n). Note that we terminate the `Table` command with a semicolon, which suppresses *Mathematica*'s output. This can be useful when the lists are long or contain large expressions. However, such a technique also hides possible mistakes. If you want to ensure that your input is correct, and yet avoid displaying the full list, use the `Table[...]` construct to generate the list (without displaying it) and then give the command `Short[%]`. This causes *Mathematica* to print out just the first few and last few points in the list. If there are mistakes, click back on the command that generates the points and edit it. Once you are satisfied that the points are being generated correctly, you can use the command `ListPlot` to display the points.

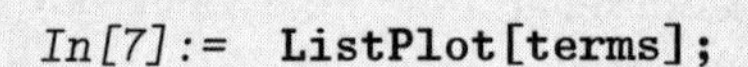

```
In[7]:= ListPlot[terms];
```

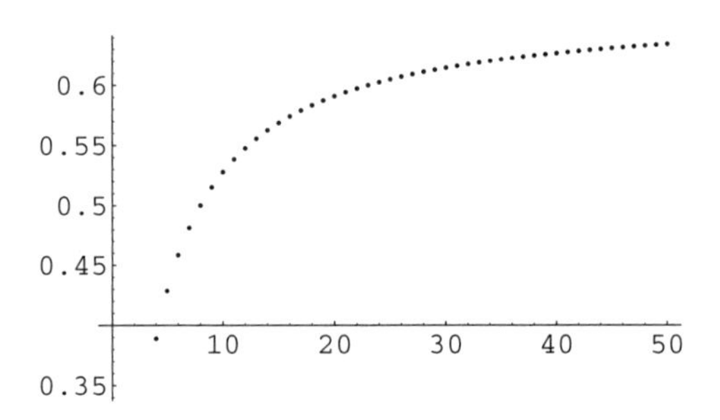

Mathematica's output is a plot of the first 50 terms of the sequence $\{b_n\}$. If you specify the option setting `PlotJoined -> True`, *Mathematica* will connect the points with straight lines. From the plot, this sequence appears to have a limit as $n \to \infty$.

```
In[8]:= Limit[b[n], n -> Infinity]

Out[8]= 2
        -
        3
```

Recall that the `Limit` command computes the limiting value of an expression.

This expression is so simple that you do not need *Mathematica* to compute the limit. However, you can use this same procedure to handle more complicated problems.

10.2 Series

Adding up some or all of the terms of a sequence is another common operation. To sum the terms, use *Mathematica*'s `Sum` command. If we let $a_n = 1/n^2$, as in Section 10.1, then

```
In[9]:=  Clear[a, n]
         a[n_] := 1/n^2
         Sum[a[n], {n, 5, 9}]

Out[11]= 737641
         -------
         6350400
```

After loading the package `Algebra`SymbolicSum``, you can use `Sum` to evaluate sums with symbolic limits, as well as limits involving infinity. Let's try this command on the infinite series $\sum_{n=1}^{\infty} \frac{1}{n^2}$.

```
In[12]:=  Needs["Algebra`SymbolicSum`"]
          Sum[a[n], {n, 1, Infinity}]

Out[13]=    2
          Pi
          ---
           6
```

Mathematica may also be able to determine that a series diverges.

```
In[14]:=  Clear[a]
          a[n_] := 1/n
          Sum[a[n], {n, 1, Infinity}]

Out[16]=  Infinity
```

Let's try one more.

```
In[17]:=  Clear[a]
          a[n_] := r^n
          Sum[a[n], {n, 0, Infinity}]

Out[19]=    1
          -----
          1 - r
```

Unfortunately, *Mathematica* did not warn us that this series converges to the value shown only if $|r| < 1$. You might try to compute the value of the series $\sum_{n=0}^{\infty} 2^n$.

Even though *Mathematica* computes the preceding examples correctly, the sum of most infinite series cannot be computed exactly. For many series, the goal is to discover whether the series converges and to compute approximate values for its sum when it does converge.

10.3 Convergence of a Series

There are four useful tests for the convergence of a series of positive terms: limit comparison, ratio, root, and integral. In this section, we use *Mathematica* and these tests to determine whether a series of positive terms converges. Consider the series

$$\sum_{n=0}^{\infty} \frac{2^{3n}}{(2n+1)!}$$

```
In[20]:=  a[n_] := 2^(3n)/(2n + 1)!
```

Again, the general term is entered as a function of `n`. This allows us to refer back to individual terms as `a[n]` etc. when we use the ratio or root test.

```
In[21]:=  Limit[a[n+1]/a[n], n -> Infinity]

Out[21]=  0
```

Since the limit of the ratio of consecutive terms is less than 1, the series converges. The same result is obtained when the limit of the nth root of the terms is computed.

```
In[22]:=  Limit[a[n]^(1/n), n -> Infinity]

Out[22]=  0
```

When the limit in the ratio or root test is different from 1, we know whether the series converges. When this limit is 1, another test is needed. For example, in the series $\sum_{n=2}^{\infty} \ln(n)/n^2$ the limit of the ratio and the limit of the nth root is 1. That is, redefining `a[n]` and applying the ratio and root tests yields:

```
In[23]:=  Clear[a ,n];
          a[n_] := Log[n]/n^2;
          {Limit[a[n + 1]/a[n], n -> Infinity],
           Limit[a[n]^(1/n),  n -> Infinity]}

Out[25]=  {1, 1}
```

For this series, because the terms form a decreasing sequence of positive terms, we can try the integral test. If we are clever, we might also compare it with the p-series where $p = 3/2$.

```
In[26]:=  a[n_] := Log[n]/n^2
          b[n_] := 1/n^(3/2)
          Limit[a[n]/b[n], n -> Infinity]

Out[28]=  0
```

Because this p-series converges $(3/2 > 1)$ and the limit of the ratios of the first series to the p-series is finite, we know that the series $\sum_{n=2}^{\infty} \ln(n)/n^2$ converges. Let's try the integral test on this series.

```
In[29]:=  Integrate[a[n], {n, 1, Infinity}]

Out[29]=  1
```

Because the integral is finite, we have a second verification that the series $\sum_{n=2}^{\infty} \ln(n)/n^2$ converges.

10.4 Error Estimates

One of the more useful aspects of the integral test is its ability to estimate the error involved in using a partial sum to approximate a series. For example, we saw earlier that the sum of the series $\sum_{n=1}^{\infty} 1/n^2$ is $\pi^2/6$. Suppose that we did not know this and added up the first 50 terms of the series. How close would we be to the actual sum?

```
In[30]:=  Clear[a, n]
          a[n_] := 1/n^2
          Sum[a[n], {n, 1, 50}]

Out[32]=  3121579929551692678469635660835626209661709
          -------------------------------------------
          1920815367859463099600511526151929560192000

In[33]:=  s50 = N[%]

Out[33]=  1.62513

In[34]:=  error = N[Pi^2/6 - s50]

Out[34]=  0.0198013
```

Now let's use the ideas underlying the integral test

$$\int_{51}^{\infty} \frac{dx}{x^2} \leq \sum_{n=51}^{\infty} \frac{1}{n^2} \leq \int_{50}^{\infty} \frac{dx}{x^2}$$

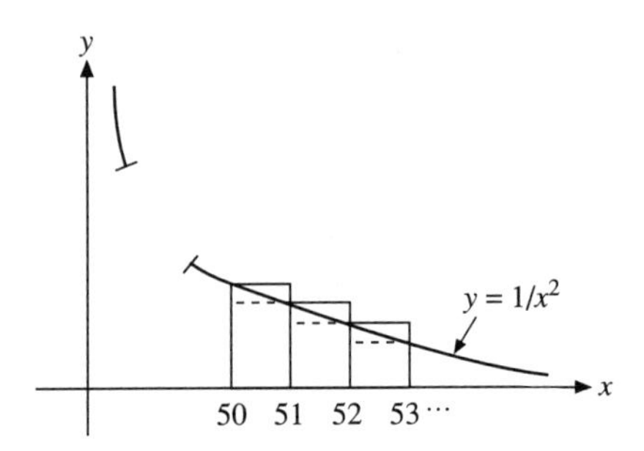


```
In[35]:= l = Integrate[1/x^2, {x, 51, Infinity}] // N

Out[35]= 0.0196078

In[36]:= u = Integrate[1/x^2, {x, 50, Infinity}] // N

Out[36]= 0.02
```

Thus, we have

$$lower = l + S50 \leq \sum_{n=1}^{\infty} \frac{1}{n^2} \leq u + S50 = upper$$

Hence, we can use the midpoint of the interval [*lower, upper*] as an approximate value for the sum of the series. The error of this approximation is at most $(upper - lower)/2$.

```
In[37]:= lower = l + s50

Out[37]= 1.64474

In[38]:= upper = u + s50

Out[38]= 1.64513

In[39]:= midPoint = (lower + upper)/2

Out[39]= 1.64494

In[40]:= maxError = (upper - lower)/2

Out[40]= 0.000196078
```

From these calculations, we see that $\pi^2/6 \approx 1.64494$.
Note: This averaging method gives us a much better approximation of the infinite series than that obtained by just using the partial sum of the first 50 terms. Moreover, this averaging method requires fewer calculations.

10.5 Taylor Polynomials

In mathematics, it is sometimes useful to approximate complicated functions with simpler ones. Polynomials are simpler than transcendental expressions, i.e., expressions that involve trigonometric and logarithmic terms such as $\sin x$, $\log x$, and e^x.

Suppose that we want the fifth-degree Taylor polynomial of $\sin(x)$. The following *Mathematica* commands construct this polynomial.

```
In[41]:= Series[Sin[x], {x, 0, 6}]

Out[41]=       3     5
              x     x           7
         x - -- + --- + O[x]
              6    120

In[42]:= p = Normal[%]

Out[42]=       3     5
              x     x
         x - -- + ---
              6    120
```

Note that the *Mathematica* command `Series` has four parameters: the expression whose Taylor polynomial we want; the variable used in the expression; the point that we expand about; and an integer. This last parameter—in this case 6—is one more than the degree of the desired Taylor polynomial. In *Mathematica*, this number refers to the order of the error of the approximating polynomial. The result returned by `Series` is a special *Mathematica* form that includes an error term of the form `O[`*var*^{n}`]`. Use the command `Normal` to confrm this special form, a power series approximation, into a polynomial.

Taylor polynomials are particularly useful. Not only do they give an effective way to approximate a function, but the form of the error term can also often lead to valuable estimates on how well we have approximated our function. For example, suppose that we want to approximate the sine function on the interval $[0, \pi]$ with its Taylor polynomial of degree 7, and we wish to know how good an approximation we have. First, we decide to expand about the midpoint of the interval.

```
In[43]:=  Series[Sin[x], {x, Pi/2, 7}]

Out[43]=        -Pi      2      -Pi      4      -Pi      6
               (--- + x)       (--- + x)       (--- + x)
                 2               2               2
          1 - ---------- + ---------- - ---------- +
                  2              24              720

               -Pi      8
             O[--- + x]
                2

In[44]:=  p = Normal[%]

Out[44]=        -Pi      2      -Pi      4      -Pi      6
               (--- + x)       (--- + x)       (--- + x)
                 2               2               2
          1 - ---------- + ---------- - ----------
                  2              24              720
```

The formula for the remainder,

$$\frac{f^{(n+1)}(\xi)}{(n+1)!}(x-\pi/2)^{(n+1)}$$

involves in this case the eighth derivative of the sine function, evaluated at some point ξ between x and $\pi/2$.

```
In[45]:=  D[Sin[x],{x, 8}]

Out[45]=  Sin[x]

In[46]:=  % /. x -> xi

Out[46]=  Sin[xi]
```

We know that the absolute value of sine is never larger than 1. Moreover, the distance between x and $\pi/2$ is always less than or equal to $\pi/2$ (recall that x lies in the interval $[0, \pi]$). Since we are expanding about the point $\pi/2$, we can therefore estimate the error as follows:

```
In[47]:=  N[(Pi/2)^8/8!]

Out[47]=  0.00091926
```

Thus, our seventh-degree polynomial is uniformly within 0.00092 of the value of the sine function on the interval $[0, \pi]$. To visualize this, we plot the sine function and the Taylor polynomial together.

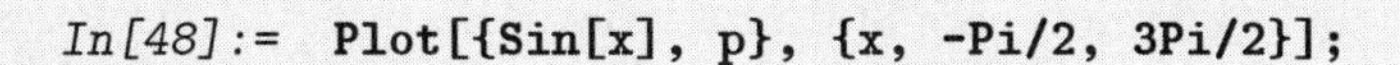

```
In[48]:=  Plot[{Sin[x], p}, {x, -Pi/2, 3Pi/2}];
```

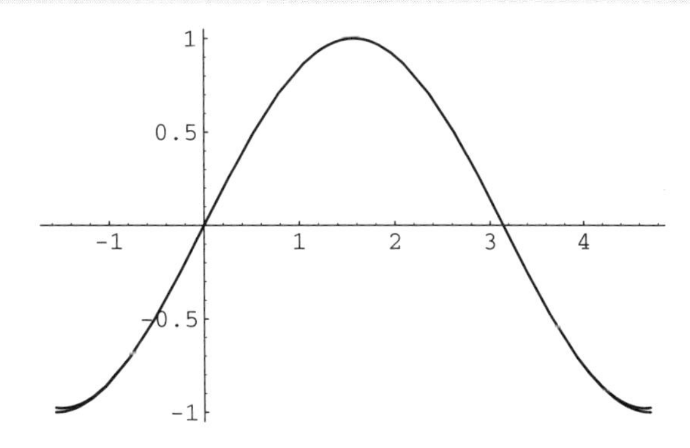

10.6 Summary

The *Mathematica* commands that we saw in this chapter are `Table`, `Sum`, `Series`, and `Normal`. Remember that you need to load the package ``Algebra`SymbolicSum` `` if you want *Mathematica* to evaluate symbolic sums, i.e., sums with non-numerical limits, in closed form. You should know what the syntax is for calling these commands. Finally, you should know how to plot sequences.

10.7 Exercises

1. Plot each of the following sequences. Try to determine whether the sequence has a limit as $n \to \infty$ and what the limiting value is. Then, have *Mathematica* compute each limit that exists.

   (a) $a_n = \dfrac{2 + (-1)^n n^2}{n^2 - 3n + 4}$

   (b) $a_n = \dfrac{\ln(n)}{n^{1/3}}$

   (c) $a_n = \dfrac{n!}{200^n}$

   (d) $a_n = \dfrac{2n^3 - 6n^2 + 15}{n^4 + 18n^3 - 6}$

   (e) $a_n = (\cos(3n))^{1/n}$

2. Sum each of the following series using *Mathematica*. Then, simultaneously plot the sequence and the sequence of partial sums for each of the series.

   (a) $\sum_{n=1}^{\infty}(-1)^n$

   (b) $\sum_{n=1}^{\infty}\frac{(-1)^{n-1}}{n}$

   (c) $\sum_{n=1}^{\infty}\frac{1}{n^k}$, for various values of k

   (d) $\sum_{n=1}^{\infty}\frac{n^3}{2^n}$

   (e) $\sum_{n=1}^{\infty}\frac{1}{1+n^2}$

3. Decide whether each of the following series converges. Try as many of the convergence tests discussed in this chapter as seem applicable.

   (a) $\sum_{n=1}^{\infty}\frac{n^2}{\sqrt{n^5+n^2+2}}$

   (b) $\sum_{n=1}^{\infty}\frac{n}{\ln\left(n^n\right)}$

4. Decide whether each of the following series converges. Give the values of x for which the series converges. Try as many of the convergence tests discussed in this chapter as seem applicable.

   (a) $\sum_{n=1}^{\infty}(\sqrt[n]{2}-1)(x-2)^n$

   (b) $\sum_{n=1}^{\infty}\frac{(-1)^n x^{2n}}{n\cos\left(n\right)}$

   (c) $\sum_{n=1}^{\infty}3^n x^n$

   (d) $\sum_{n=1}^{\infty}\frac{(-1)^n x^n}{n2^n}$

5. Find the requested Taylor polynomial and estimate the error in approximating the given function with the Taylor polynomial on the given interval. Then simultaneously plot both the function and the Taylor polynomial.

   (a) The fourth-degree Taylor polynomial of $\cos(3x)$ about $x = \pi/4$ on the interval $[0, \pi]$.

   (b) The first-, fifth-, and fifteenth-degree Taylor polynomials of

   $$\frac{x^4 - 16x^2 + 2x - 5}{x^2 - 6}$$

   about $x = 1$. Plot all of them and the given function on the same graph for the interval $[-1, 2]$.

   (c) The 100th-degree Taylor polynomial of $x^4 - 2x^2 + 15x - 6$ about $x = 0$ and also about $x = 5$, on the interval $[-2, 8]$.

6. Estimate the value of the series $\sum_{n=1}^{\infty} \frac{1}{n^3}$ by using the sum of the first 25 terms of the series and the averaging method discussed in Section 10.4.

7. Expand the function $\sin(x)$ about $x = \pi/2$ in a Taylor polynomial of degree six and also in one of degree seven. The seventh-degree polynomial should be more accurate that the sixth. Is it? How much extra work is involved in evaluating this seventh-degree Taylor polynomial than in evaluating the sixth-degree one?

11 Differential Equations

In this chapter, we demonstrate how to use *Mathematica* to solve first-order and second-order differential equations. The appropriate *Mathematica* command is `DSolve`. *Mathematica* is good at finding explicit solutions for differential equations—when they can be found. Unfortunately, many differential equations cannot be solved using standard techniques. In such a case, you either have to write your own differential equation solver that uses a specialized technique that you believe will work, or you have to settle for a numerical solution. The latter is obtained using the *Mathematica* command `NDSolve`. We will show you how to call both `DSolve` and `NDSolve` and how to work with the solutions they return.

11.1 Explicit Solutions

To solve the differential equation $y' + 5y = 2t$, input

```
In[1]:=  eq1 = (y'[t] + 5 y[t] == 2 t);
         DSolve[eq1, y[t], t]

Out[2]=                2      2 t    C[1]
         {{y[t] -> -(--) + --- + ----}}
                      25      5      5 t
                                   E
```

Note that *Mathematica* gives the answer in the form of a replacement rule for $y(t)$. Also notice the term `C[1]`. This term is *Mathematica*'s notation for the arbitrary constant that occurs in the general solution of a first-order differential equation.

We specify the initial condition, $y(1) = 2$, as another equation.

```
In[3]:=  DSolve[{eq1, y[1] == 2}, y[t], t]

Out[3]=                          5 - 5 t
                       2      42 E            2 t
         {{y[t] -> -(--) + ----------- + ---}}
                      25        25          5
```

Note that, in the input command, the equation and the initial condition are enclosed in braces. To see a plot of the solution, use the *Mathematica* command `Plot[y[t] /. %, {t, 0, 3}]`. The command `/.` is a shorthand notation for `ReplaceAll` and `%` refers to the last output *Mathematica* calculated. The construct `y[t] /. %` therefore causes the value of `y[t]` to be replaced with the right-hand side of the rule given as the solution to the differential equation.

The commands for solving a second-order (or higher-order) differential equation are similar. To find the general solution to the second-order differential equation

$$\frac{d^2y}{dt^2} + 6\frac{dy}{dt} - 7y = 1$$

enter the commands

```
In[4]:=  eq2 = (y''[t] + 6 y'[t] - 7 y[t] == 1);
         DSolve[eq2, y[t], t]

Out[5]=              1     C[1]      t
         {{y[t] -> -(-) + ---- + E  C[2]}}
                     7     7 t
                          E
```

Note that there are two arbitrary constants in the solution, `C[1]` and `C[2]`. Their values depend on the particular initial conditions or boundary conditions associated with the differential equation. By specifying such conditions, you can have *Mathematica* determine a unique solution to the differential equation. For example, if $y(1) = 3$ and $y'(1) = 2$, you would enter:

```
In[6]:=  DSolve[{eq2, y[1] == 3, y'[1] == 2}, y[t], t]

Out[6]=                   7 - 7 t
                     1   E                -1 + t
         {{y[t] -> -(-) + -------- + 3 E        }}
                     7       7
```

The solution now contains numerical values for `C[1]` and `C[2]`.

11.2 Direction Fields

Graphing the direction field associated with the differential equation can help you understand the behavior of the solutions to a first-order differential equation.

Suppose that we want to investigate the solutions to the differential equation $y' = x\sin(y)$ by first examining its direction field. The command `DirectionField` graphs direction fields. For convenience, we bundle the necessary operations in a simple function that implements the calculations needed to define the direction field:

```
In[7]:= DirectionField[Derivative[1][y_][x_] == f_,
                xRange_, yRange_,n_:17]:=
            DirectionField[f /. y[x]->y, xRange, yRange, n]

        DirectionField[f_, {x_,a_,b_}, {y_,c_,d_}, n_:17]:=
            Module[{dx,dy,l},
                l = Sqrt[(b - a)^2 + (d - c)^2]/(2 n);
                dx = N[(b - a)/(n - 1)];
                dy = N[(d - c)/(n - 1)];
                Show[
                    Graphics[
                        Flatten[
                            {RGBColor[1,0,0],
                             PointSize[0.01],
                             Table[
                                 MakeLine[f/.{x->xi,y->yi},
                                     xi, yi, l],
                                 {xi,a,b,dx}, {yi,c,d,dy}]
                        }],
                     Axes->True,
                     AxesLabel->{x,y}
                     ]
                ]
            ]


         (* If the gradient is indeterminate, do not
          * attempt to draw it!
          *)
        MakeLine[Indeterminate, xi_, yi_, l_]:=
            {Point[{xi,yi}]}

        (* When Mathematica generates an expression
         * involving Infinity, instead use gradient.
         *)
        MakeLine[gradient_, xi_, yi_, l_]:=
            {Point[{xi,yi}],
             Line[{{xi, yi - 0.5 l},{xi, yi + 0.5 l}}]} /;
                MemberQ[{ComplexInfinity, -ComplexInfinity,
                         Infinity, -Infinity},
             gradient]

        MakeLine[gradient_, xi_, yi_, l_]:=
            Module[{theta},
                theta = ArcTan[gradient];
                xOffset = 0.5 l Cos[theta];
                yOffset = 0.5 l Sin[theta];
                {Point[{xi,yi}],
                 Line[{{xi - xOffset, yi - yOffset},
                       {xi + xOffset, yi + yOffset}}
                 ]
                }
             ]
```

Henceforth, we can use this function to produce a plot of the direction field. For example:

```
In[12]:=  DirectionField[x Sin[y], {x,-2,2}, {y,-1,1}]
```

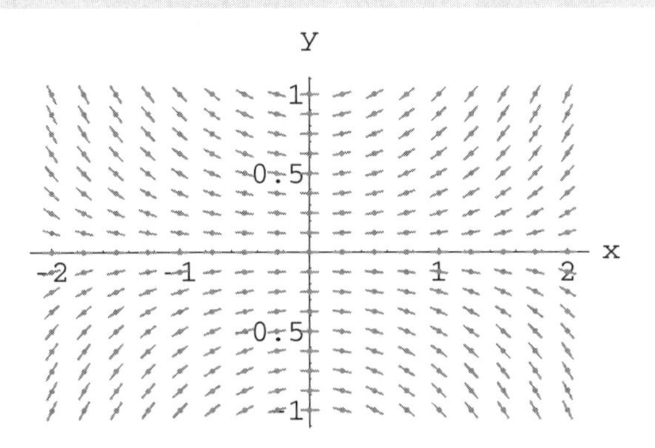

To graph a solution to the differential equation $y' = x\sin(y)$ on top of the direction field, use the following procedure. First, solve the differential equation, plot the direction field associated with the differential equation, and then superimpose both plots using the `Show` command. In the case of the equation $y' = x\sin(y)$, the differential equation solving techniques that come preloaded into *Mathematica* are not powerful enough to solve the equation. For example, if you give the command:

```
In[13]:=  DSolve[{y'[x] == x Sin[y[x]], y[0] == 0.25},
             y[x], x]
```

you get a warning message and the input is returned to you unevaluated:

```
Solve::ifun:
   Warning: Inverse functions are being used by
     Solve, so some solutions may not be found.
Out[13]=  DSolve[{y'[x] == x Sin[y[x]], y[0] == 0.25},
             y[x], x]
```

Fortunately, there is a package that enlarges the scope of the differential equations that *Mathematica* can solve. Load the package ``Calculus`DSolve` `` with the command ``Needs["Calculus`DSolve`"]``. Then, try the same differential equation again:

```
In[14]:=  Needs["Calculus`DSolve`"]
          DSolve[{y'[x] == x Sin[y[x]], y[0] == 0.25},
              y[x], x]

Solve::ifun:
   Warning: Inverse functions are being used by
     Solve, so some solutions may not be found.
Solve::ifun:
   Warning: Inverse functions are being used by
     Solve, so some solutions may not be found.

Out[14]=                                                 2
                              0.5(-4.14842827160637 + x )
          {{y[x] -> 2. ArcTan[E                            ]}}
```

In this case, although *Mathematica* prints warning messages, the differential equation is solved. Putting it all together, if we want to plot the solution of the differential equation together with the direction field, we type:

```
In[15]:=  Needs["Calculus`DSolve`"];
          plot1 = DirectionField[x Sin[y],
              {x, -2, 2}, {y, -1, 1}];
          ysol = DSolve[{y'[x]==x Sin[y[x]], y[0]==0.25},
              y[x], x];
          plot2 = Plot[y[x] /. ysol, {x, -2, 2}];
          Show[plot1, plot2]
```

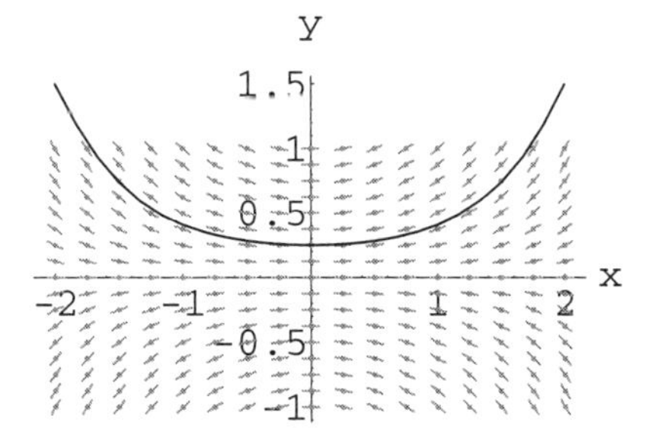

Notice how the solution follows the direction field. It does so because the tangent line to the solution is parallel to the line given by the direction field. In fact, it is this idea that gives rise to an easily implemented numerical algorithm for calculating approximate solutions to differential equations called "Euler's Method."

11.3 Numerical Solutions

Mathematica can find numerical approximations to the solutions of differential equations. For example, suppose that we want to solve $y' + \sin(y^2) = 1,\ y(0) = 1$. We can obtain a numerical solution using the *Mathematica* command `NDSolve`. In general, whenever you want to solve a differential equation numerically, you must specify enough initial conditions or boundary conditions that the solution can be determined uniquely. As $y' + \sin(y^2) = 1,\ y(0) = 1$ is a first-order equation, a single initial condition is sufficient. Moreover, when solving a differential equation numerically, it is also necessary to specify the range of the independent variable over which a solution is sought. So if we want the solution over the interval $[0, 4]$, we could enter the *Mathematica* command:

```
In[20]:= ndSolution =
             NDSolve[{y'[t]+Sin[y[t]^2] == 1, y[0] == 1},
                 y[t], {t, 0, 4}]

Out[20]= {{y[t] -> InterpolatingFunction[{0., 4.}, <>][t]}}
```

Note that *Mathematica* outputs the answer (i.e., the numerical solution to the differential equation) in terms of `InterpolatingFunction`. The full form is suppressed, as indicated by the characters <>, because users probably would not find it useful to view the interpolation data directly.

Notice also that the answer is enclosed inside two layers of lists. The reason is that, for some initial conditions, there might be more than one solution. In such cases, *Mathematica* would return a list of the solutions. This explains the outer list. Moreover, if, instead of solving a single differential equation, we had used `NDSolve` to solve a system of differential equations, then each solution would be specified as a list of `InterpolatingFunction` objects, one for each dependent variable. This explains the inner list.

The actual meat of the solution is the `InterpolatingFunction` object on the right-hand side of the replacement rule for `y[t]`. This object can be manipulated as though it were a regular *Mathematica* expression. For example, if we want to use the `InterpolatingFunction` object to find approximate values for the solution to the differential equation at various values of t, e.g., $t = 3.1$, the easiest way to do so is with the commands

```
In[21]:= y[t] /. ndSolution /. t -> 3.1

Out[21]= {1.17382}
```

Similarly, we can graph the solution over the specified interval.

```
In[22]:= Plot[y[t] /. ndSolution, {t, 0, 4}];
```

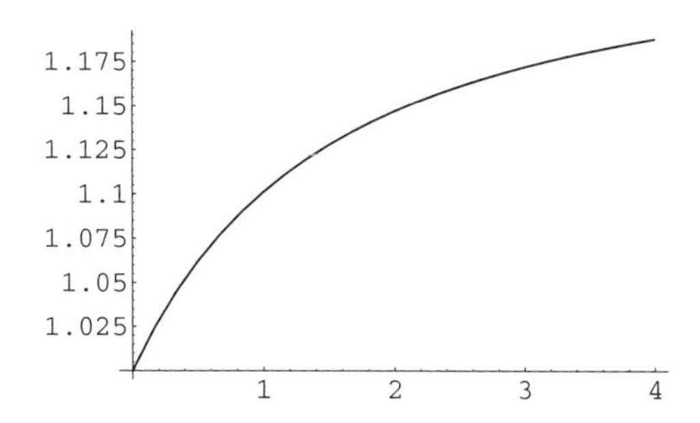

11.4 Summary

You can solve a differential equation symbolically using the `DSolve` command and numerically using the `NDSolve` command. You can plot the solutions using the `Plot` command. You can display several plots simultaneously using the `Show` command.

11.5 Exercises

1. Use *Mathematica* to solve the following differential equations. If an initial value is given, plot the solution.

   (a) $\frac{dy}{dx} = \frac{xy + 3x}{x^2 + 1}$, $y(2) = 1$

   (b) $\frac{dy}{dx} = \frac{\ln(x)}{xy + xy^3}$

   (c) $y' + (\cos x)y = \cos x$

   (d) $y'' + \sin(y) = 0,\ y(0) = 1,\ y'(0) = 0$

2. Solve the equation $y' + y = \sin(x)$, with initial condition $y(0) = 0$, both numerically and exactly. Plot both solutions together in the range $-2 \leq x \leq 10$. How well does the numerical solution approximate the exact solution?

3. Plot the direction field of the differential equation $y' = x - y$. On a sheet of paper, hand sketch what you think the graph of the solution to this differential equation and the initial condition $y(0) = 0$ will look like. Then, plot the solution and the direction field together.

12 Parameterized Curves and Polar Plots

In this chapter, we will use *Mathematica* to create parametric plots and polar plots.

12.1 Parameterized Curves and Polar Plots

Some figures, such as circles and ellipses, are not the graphs of functions. Instead, these figures are more conveniently described by *parametric equations*, which are of the form

$$x = f(t) \quad y = g(t), \ \ a \leq t \leq b$$

where f and g are functions of the parameter t, and a and b are numbers. As a simple example, a circle of radius 3 centered at the origin is described by the parametric equations

$$x = 3\cos(t) \quad y = 3\sin(t), \ \ 0 \leq t \leq 2\pi.$$

Parametric equations are often used to describe the position of a particle at time t.

To plot the parametric equations

$$x = f(t) \quad y = g(t), \ \ a \leq t \leq b$$

we use the *Mathematica* function `ParametricPlot`. If f and g are expressions in t, rather than functions of t, the syntax is the same, except that f and g would be entered, rather than $f(t)$ and $g(t)$.

Example 12.1 To plot a circle of radius 3 centered at the origin, type

```
In[1]:=  ParametricPlot[{3 Cos[t], 3 Sin[t]}, {t, 0, 2 Pi}];
```

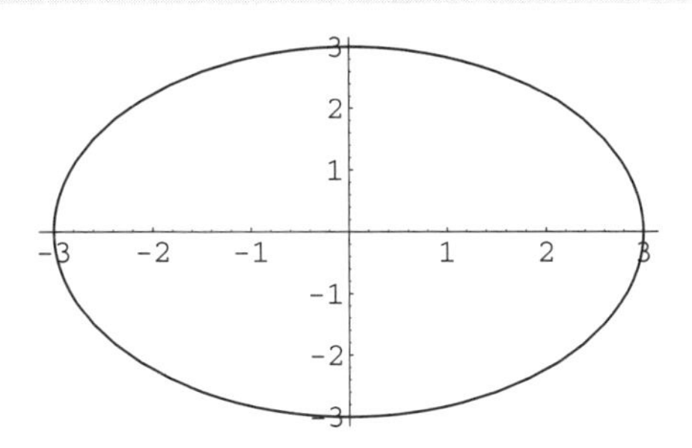

Your circle may look deformed (more like an ellipse), because the scales for the x- and y-axes are different. That is, one unit in the x-direction is not drawn the same size as one unit in the y-direction. To make the curve look like a circle, instead of an ellipse, specify the value `Automatic` for the option `AspectRatio` so that one unit in the x-direction is drawn the same size as one unit in the y-direction.

```
In[2]:=  ParametricPlot[{3 Cos[t], 3 Sin[t]}, {t, 0, 2 Pi},
             AspectRatio->Automatic];
```

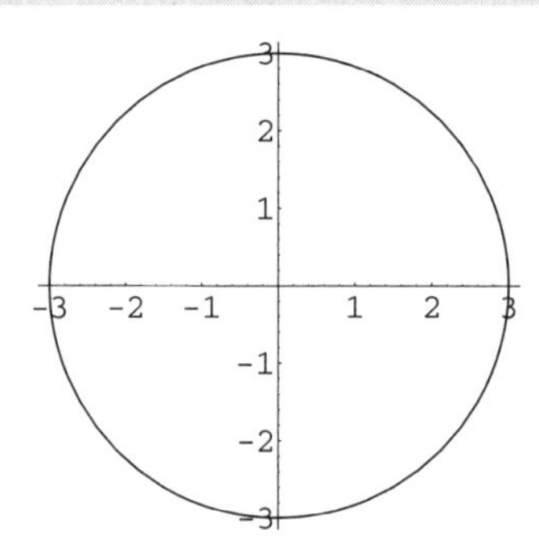

Example 12.2 Plot the polar equation

$$r = \cos(3\theta), \quad 0 \le \theta \le 2\pi$$

Solution. Recall that a polar curve is an example of a parametric curve with the parameter θ playing the role of t. Since it is easier to type t than to type θ, we will use t for the parameter. To plot the polar equation $r = \cos(3t)$, convert from polar coordinates to rectangular coordinates using

$$x = r\cos(t) \qquad y = r\sin(t)$$

and enter the commands

```
In[3]:=  r = Cos[3 t];
         ParametricPlot[{r Cos[t], r Sin[t]}, {t,0,2Pi},
             AspectRatio->Automatic];
```

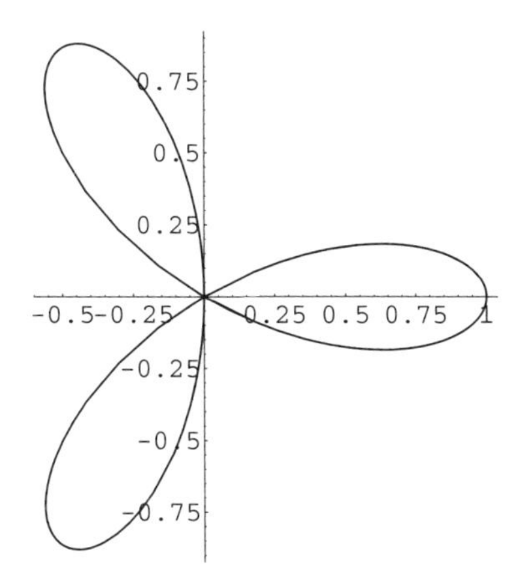

Example 12.3 Plot the parabola given in polar form by $r = 1/(1 - \sin(t))$.
Solution. First, enter the expression for r; then, enter the rectangular coordinate expressions for this parabola.

```
In[5]:=  r = 1/(1-Sin[t]); x = r Cos[t]; y = r Sin[t];
         ParametricPlot[{x, y}, {t, 0,  Pi},
              AspectRatio -> Automatic];
```

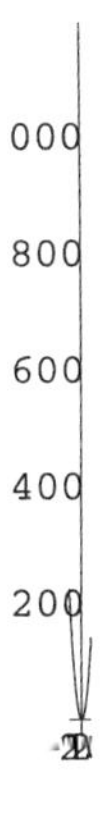

Note that the scale of the plot is large because the expression r is large for t near $\pi/2$. To view a portion of the graph near the origin, specify the value {{-5, 5}, {-1, 5}} for the option `PlotRange`, as follows:

```
In[7]:=  r = 1/(1-Sin[t]); x = r Cos[t]; y = r Sin[t];
         ParametricPlot[{x, y}, {t, 0, 2 Pi},
              PlotRange -> {{-5,5}, {-1,5}},
              AspectRatio -> Automatic];
```

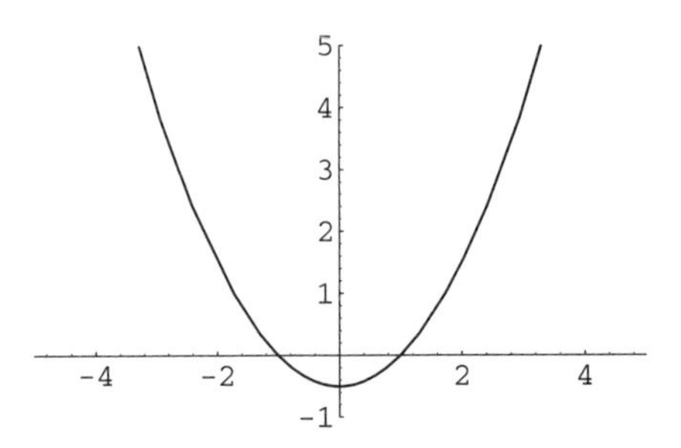

Alternatively, you can use the command `PolarPlot` that is designed specifically for producing polar coordinate plots. It is defined in the package `Graphics`Graphics``. Make sure to load this package before calling `PolarPlot`. Then, you can plot the parabola using `PolarPlot`. To specify the ranges $-5 \leq x \leq 5$ and $-1 \leq y \leq 5$, you must use the `PlotRange` option by entering

```
In[9]:=  Needs["Graphics`Graphics`"];
         PolarPlot[r, {t, 0, 2Pi},
            PlotRange -> {{-5,5}, {-1,5}}];
```

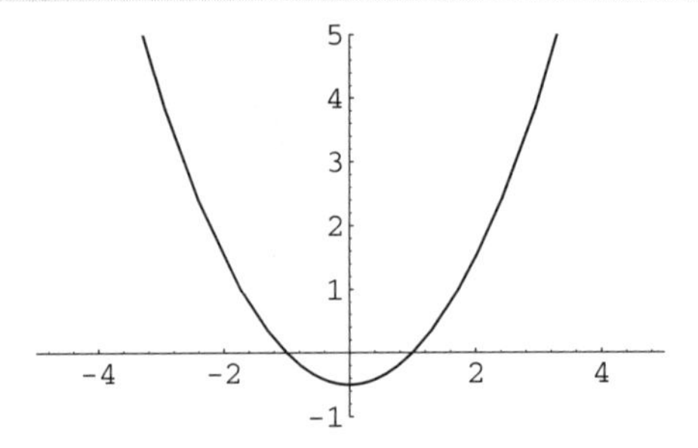

12.2 Summary

You can use the command `ParametricPlot` to create parametric and polar plots. In addition, you can use the specialized command `PolarPlot` to create polar plots. To use the latter command, you must first load the package `Graphics`Graphics``.

12.3 Exercises

1. Plot the ellipse $x = 3\cos(t)$ and $y = \sin(t)$ using constrained scaling.
2. Plot a piece of the hyperbola parameterized by $x = \cosh(t)$ and $y = \sinh(t)$, for $-2 \leq t \leq 2$. Recall that $\cosh(t) = (e^t + e^{-t})/2$ and $\sinh(t) = (e^t - e^{-t})/2$.

3. Graph each of the following equations by hand, and then check your answers with a *Mathematica* plot. If the graph is a conic section (ellipse, hyperbola, or parabola), then give the location of the foci or focus.

   (a) $x^2 + 2y^2 = 4$

   (b) $r = \dfrac{3}{1 - \cos(t)}$

   (c) $r = \dfrac{2}{1 - 2\sin(t)}$

   (d) $r = \dfrac{4}{1 + 2\cos(t)}$

   (e) $r = \dfrac{4}{1 - \sin(t + \pi/3)}$

   (f) $r = \dfrac{2}{3 + \cos(t + \pi/4)}$

   (g) $r = 2 + \cos(t)$

4. Recall that the formula for computing the area of the region that lies between a polar graph $r = r(t)$ and the origin, and between the angles $t = a$ and $t = b$, is given by

$$\frac{1}{2}\int_a^b r(t)^2\, dt$$

   Set up and evaluate the integral(s) required to find the area that is inside both the parabola $r1(t) = 3/(1 - \cos(t))$and the circle $r2(t) = 8$. *Hint:* First, plot these two graphs to estimate the angles where the two curves cross. Next, use `NSolve` to find the angles. Then, perform the appropriate integrals using *Mathematica*'s `NIntegrate`.

5. Recall that the arc length of a polar graph $r = r(t)$, $a \le t \le b$, is given by

$$\int_a^b \sqrt{r(t)^2 + r'(t)^2}\, dt$$

   The formulae for finding the surface area of revolution of this polar graph about the x- and y-axes are, respectively

$$2\pi \int_a^b y(t)\sqrt{r(t)^2 + r'(t)^2}\, dt$$

$$2\pi \int_a^b x(t)\sqrt{r(t)^2 + r'(t)^2}\, dt$$

   provided that a and b are appropriate (see exercise 6). Here, $x(t) = r(t)\cos(t)$ and $y(t) = r(t)\sin(t)$.

(a) Set up and evaluate the integral involved in calculating the arc length of the curve $r = 2 + \cos(t),\ 0 \leq t \leq \pi/2$.

(b) Set up and evaluate the integral involved in calculating the area of the surface that you obtain by revolving the curve $r = 2 + \cos(t),\ 0 \leq t \leq \pi/2$ about the x-axis.

(c) Set up and evaluate the integral involved in calculating the area of the surface that you obtain by revolving the curve $r = 2 + \cos(t),\ 0 \leq t \leq \pi/2$ about the y-axis.

6. Set up and evaluate the integral involved in calculating the area of the surface that you obtain by revolving the petal of the rose $r = \cos(3t),\ -\pi/6 \leq t \leq \pi/6$, about the x-axis. Be careful in choosing your limits of integration.

13 Programming with *Mathematica*

We have indicated in earlier chapters that *Mathematica* has a built-in programming language. In fact, it is possible to use *Mathematica* to create highly structured programs that can perform a large number of useful tasks. Any programs written in *Mathematica* can call any of the high-level packages available in *Mathematica*, or any programs that you have written in *Mathematica* and included in the same session. This chapter provides an introduction to the *Mathematica* programming environment. It will familiarize you with just enough details of *Mathematica*'s programming capabilities to let you write your own calculus-oriented programs. We assume that you have had some previous experience with structured programming.

13.1 Introduction to *Mathematica* Programs

Mathematica programs are defined in a way similar to *Mathematica* functions. Unlike functions, however, a program may produce several outputs and have various side effects, such as resetting the value of global variables or printing data to a file. Programs may call user-defined functions or functions available in *Mathematica* packages once the packages have been loaded (using the command `Needs["PackageType`PackageFile`"]`).

Most *Mathematica* programs have the structure

```
programName[arg1_, arg2_, ...] :=
     Module[{v1, v2,...},
          v1 = expr1;
          v2 = expr2;
          ...
     ]
```

where *programName* is some sequence characters and numbers, beginning with a character, that is intended to convey the purpose of the program. It is worth adhering to the naming conventions used for *Mathematica*'s built-in functions: choose meaningful names that are complete words, and, if the name contains multiple words, capitalize the first letter of each word. The inputs to the program, *arg1*, *arg2*, and so on, can be any valid *Mathematica* expressions, such as explicit numbers, symbols, lists, or matrices, or perhaps the name of a file from which data are to be read. It is possible to define programs that test whether an argument is of the right type or whether it matches a certain pattern before attempting to run the program. We shall see examples of such patterns later. Finally, the body of a *Mathematica* program lies to the right of the := and is usually a compound *Mathematica* expression–i.e., a sequence of *Mathematica* expressions separated by semicolons (;). In most non-trivial programs, you will want to introduce new variables, *v1*, *v2*, and so on, to act as temporary names for intermediate data structures. You can shield the values assigned to these variables from other functions or programs by calling `Module` with a list of the local variables as the first argument and the body of the program as the second argument. However, it is not necessary to define *Mathematica* functions in this manner. In particular, you can avoid introducing intermediate variables to hold data structures simply by nesting one *Mathematica* function within another. For example,

```
f[x_] := Sin[Cos[1-x^2]]
f[x_] := Module[{y, z},
            y = 1 - x^2;
            z = Cos[y];
            Sin[z]]
```

both define exactly the same function. In general, the nested approach is more efficient. However, if your function uses the value of some intermediate variable more than once, it is usually more efficient, computationally, to do the calculation once and to assign the result to a local variable so that it is available for reuse with no extra computation.

```
g[x_] := Sin[Cos[1-x^2]] + Tan[Cos[1-x^2]]
g[x_] := Module[{y},
            y = Cos[1-x^2];
            Sin[y] + Tan[y]]
```

If you are nesting large expressions, use indentation to lay out your program. Judicious indentation will help you understand and debug your own code. Also, you will find it helpful to annotate your program with comments. Enter these as text between the special characters (* and *).

Now let's create a few procedures that are slightly more complex than the simple example given above. The first example illustrates use of the `Table` command for building lists and the `D` function for performing symbolic differentiation.

Example 13.1 Suppose that you need a simple procedure that can be used to determine whether a pattern emerges when higher-order derivatives of a function are computed. We give an example procedure that can be used to list the first k derivatives of a given function f.

```
In[1]:=  DerivativePattern[function_, var_, k_] :=
             ColumnForm[Table[D[function, {var, n}],
                                 {n, 1, k, 1}]]
```

The `Table` command takes two arguments: an expression, involving an index variable i, and an iterator, of the form $\{i,\ i_{\mathrm{start}},\ i_{\mathrm{stop}},\ i_{\mathrm{incr}}\}$. In the program, the iterator generates a sequence of values for i from 1 to k, in increments of 1, and the expression is evaluated for each of these values in turn. There are more compact ways of specifying this particular iterator, but we want to show you the most general version of an iterator so that you will be able to handle more complicated cases in your own programs. The *Mathematica* function `D[`*function*`,{`*var*`,`*n*`}]` means differentiate *function* with respect to the variable *var* *n* times. So, putting it all together, our new function `DerivativePattern` takes some expression, the variable with respect to which to differentiate, and the highest derivative desired and returns a list of all derivatives up to the highest one desired, laid out in a column on the screen.

Now let's call our new program on the expression `Exp[x] Sin[x]` to see whether we can spot any pattern.

```
In[2]:=  DerivativePattern[Exp[x] Sin[x], x, 5]

Out[2]=   x          x
         E  Cos[x] + E  Sin[x]

            x
         2 E  Cos[x]

            x            x
         2 E  Cos[x] - 2 E  Sin[x]

             x
         -4 E  Sin[x]

             x             x
         -4 E  Cos[x] - 4 E  Sin[x]
```

Observe that the fourth derivative of the function is -4 times the function. Similarly, we see that 2 times the first derivative of the function minus the second derivative of the function is 2 times the function. We might try using this information to write a variety of differential equations that the function $f(x) = e^x \sin(x)$ satisfies.

How can we make *Mathematica* do more of the work? In particular, can we make *Mathematica* spot a pattern for us? There are two useful functions that will help: `MatchQ` and `Select`. We can find out about these functions using the `?` operator.

```
In[3]:=  ?MatchQ

MatchQ[expr, form] returns True if the pattern form
matches expr, and returns False otherwise.

In[4]:=  ?Select

Select[list, crit] picks out all elements ei of list for
which crit[ei] is True. Select[list, crit, n] picks out
the first n elements for which crit[ei] is True.
```

Let's use these functions to define a new version of our derivative pattern program that automatically picks out higher derivatives that are constant multiples of the original function:

```
In[5]:=  AutoDerivativePattern[function_, var_, k_]:=
           Module[{derivs, const},
                derivs = Table[{i, D[function, {var, i}]},
                     {i,1,k,1}];
                Select[derivs,
                     MatchQ[#[[2]],const_Integer function]&
                ]
           ]
```

In `AutoDerivativePattern`, `derivs` is the name given to a list of pairs, the first element of each pair being the order of the derivative, and the second being the derivative itself. The construct `Select[derivs, MatchQ[...]]` picks out just those pairs that contain derivatives that are a constant multiple of the original expression. You would call this function as follows:

```
In[6]:=  AutoDerivativePattern[Exp[x] Sin[x], x, 5]

Out[6]=              x
         {{4, -4 E  Sin[x]}}
```

This version of `AutoDerivativePattern` uses what is called a *pure function*. Pure functions are a shorthand way of achieving the effect of a function without having to define that function explicitly. For more information on pure functions see *Mathematica: A Practical Approach* by Nancy Blachman (Prentice-Hall, 1992).

Example 13.2 Suppose that you need a program that plots a function along with its tangent lines at several points. We define such a program here: its inputs are (1) the function to be plotted, (2) the interval for the plot, and (3) the list of x-coordinates at which the tangent lines are to be drawn.

```
In[7]:=  TangentPlot[fn_, {x_, a_, b_}, xValues_List]:=
             Module[{tanLines, tanPlots},
                 tanLines =
                     Map[EquationOfTangent[fn,x,#]&,
                         xValues];
                 tanPlots = Map[Plot[#,{x,a,b},
                     DisplayFunction->Identity]&,tanLines];
                 Show[Plot[fn, {x,a,b},
                     DisplayFunction->Identity],
                 tanPlots,
                 DisplayFunction->$DisplayFunction]
             ]

         EquationOfTangent[fn_, x_, x0_]:=
             N[(D[fn,x] /. x->x0) (x - x0) + (fn /. x->x0)]
```

We can use this program to plot $4\sin(x)+\cos(2x)$, along with its tangent lines at $x=-1,2,4$ on the interval $[-3,6]$ as follows:

```
In[9]:=  TangentPlot[4 Sin[x]+Cos[2 x], {x,-3,6}, {-1,2,4}]
```

In this case, we employed the local variables `tanLines` and `tanPlots` in our program. Thus, values assigned to these variables are defined only within the context of the function `TangentPlot`. Consequently, the variable names `tanLines` and `tanPlots` may be used in other functions. We also used the command `Map`.

If you are not familiar with this command, you can use `?` to find out about it.

```
In[10]:=  ?Map

Map[f, expr] or f /@ expr applies f to each element on
the first level in expr. Map[f, expr, levelspec] applies
f to parts of expr specified by levelspec.
```

The first argument of the `Map` command is a function that we want to evaluate for a finite set of different values; the second argument of the `Map` command is a list that contains those different values. `Map` peels off each value in succession, feeds that value to the function that we want to evaluate, and accumulates the results in a list.

Note that, although there are several executable statements with semicolons, we see the output only from the final one. That is because a *Mathematica* function returns the value only of the last statement executed.

We close this section with a simple illustration that *Mathematica* offers a variety of programming styles. Sometimes, one style is more natural than another for a given problem. It is the hallmark of a good *Mathematica* programmer to be able to switch styles to suit the task at hand.

Example 13.3 Suppose that you need a program that evaluates a function for several input numbers, and then sums the results. *Mathematica* offers three distinct ways of doing this according to three different programming styles: functional, rule-based, and procedural. We shall write a version of a summation program in each style as a way of illustrating the differences among them. Let's call the functional version SumProgramF, the rule-based version SumProgramR, and the procedural version SumProgramP. We will also show you how to tell which version is the most efficient.

```
In[11]:=  (* functional *)
          SumProgramF[fn_, x_, xValues_List]:=
              Apply[Plus, Map[(fn /. x->#)&, xValues]]

          (* rule-based *)
          SumProgramR[fn_, x_, xValues_List]:=
              (fn /. x->xValues) //. {p__, q_}->p+q

          (* procedural *)
          SumProgramP[fn_, x_, xValues_List]:=
              Module[{total = 0},
                  For[i=1, i<=Length[xValues], i++,
                      total += fn /. x->xValues[[i]]
                  ];
                  total
              ]
```

In each case we have used the construct `fn /. x ->` *value* to evaluate the function `fn` and to replace all occurrences of `x` with the value `xValue`. This construct can be interpreted as "Evaluate `fn` given that `x` goes to `xValue`."

The functional version `SumProgramF` uses the `Map` command that we saw earlier to map this evaluation operation down a list of desired values for x, creating a list of function values. Then, *Mathematica* sums all these function values by applying the `Plus` operation to this list.

The rule-based version `SumProgramR` relies on a second use of the rule syntax. Whereas `f /. x -> y` causes occurrences of `x` in `f` to be replaced with `y`, the syntax `f /. x -> {y1,y2,y3,...}` causes a whole list of values for `f` to be returned, corresponding to a replacement rule using each `yi` in turn. Then, the resulting list is matched against the pattern `{p__, q_}`. Whereas a variable such as `p_` (i.e., `p` with one underscore) matches any single *Mathematica* expression, a variable such as `p__` (`p` with two underscores) matches any sequence of one or more *Mathematica* expressions. Consequently, a rule such as `list /. {p__,q_}->p+q` has the effect of matching `p` to all but the last element of the list, and `q` to the last element. However, `Plus` distributes over lists, and thus we compute the sum of all the elements in the list.

Finally, the procedural version SumProgramP is merely a For loop with the total initialized to zero, and each successive step through the list of function values adds the next value to the current partial sum.

How long do these different versions take to execution? Luckily there is a built-in *Mathematica* function called Timing that returns the amount of CPU time used to compute a result. Using Timing, we can write a program that computes the time to evaluate the sum using the three versions of our summation program when they are given a large argument.

```
In[12]:=  CompareTimes[fn_, x_, xValues_]:=
                  {Timing[SumProgramF[fn,x,xValues]],
                   Timing[SumProgramR[fn,x,xValues]],
                   Timing[SumProgramP[fn,x,xValues]]}
```

We can use a list of the first 1000 integers as a suitable test data set, as follows:

```
In[13]:=  CompareTimes[x^2 + x, x, Range[1000]]

Out[13]=  {{0.95 Second, 334334000},
           {0.383333 Second, 334334000},
           {1.73333 Second, 334334000}}
```

In this case, the rule-based program was the most efficient. In *Mathematica*, functional and rule-based programs are almost always faster than procedural programs. Avoid using For, Do, and While loops whenever possible.

There are certainly numerous *Mathematica* programming structures that we have not included in this short introduction. For example, *Mathematica* has built-in If, Switch, and Which commands that permit conditional control of *Mathematica*'s execution sequence. You can also add conditions and patterns to the arguments of a function that control whether a particular clause will apply. For example, you may want a function that performs a different evaluation, depending on whether the argument is an even or an odd number. Such a function can be defined in two clauses whose heads are foo[x_/;EvenQ[x]]:= ... and foo[x_/;OddQ[x]]:=....

13.2 Creation of Interactive Programs in *Mathematica*

The example in this section illustrates how to write an interactive program in *Mathematica*. The Input command is used to obtain input from a user. When you call Input from a Notebook Front End, *Mathematica* pops open a window in which you are to type your reply. Responses to prompts created by Input must use correct *Mathematica* syntax. If the response is text, then it must be included in double quotes: e.g., "This is an answer." However, if you want only strings to be returned, you should use the specialized *Mathematica* function InputString.

Example 13.4 To illustrate the use of the `Input` command, we create a simple drill program that can be used to test your differentiation skills. We shall keep it simple to emphasize the main points. You may wish to modify this program later to make a more sophisticated self-testing program for differentiation or for some other mathematical domain. This program introduces the `Random` command (for generating pseudo-random numbers), demonstrates the use of `If` statements, and illustrates that it is possible to write programs that do not require arguments. The program asks a simple question, and then checks the response of the user. The process continues until three correct responses are given. First, let's look at the outer driver that runs the test until three correct answers have been given. The driver is implemented in the top-level function `TakeTest`, which has no arguments but must be called with square brackets because we defined it that way.

```
TakeTest[ ]:=
    Module[{numSuccesses=0},
        While[numSuccesses < 3,
            numSuccesses +=
                PracticeDifferentiation[ ]
        ];
        Print["Well done! :-) Goodbye."]
    ]
```

The function `PracticeDifferentiation` does all the work. For now, all we need to know is that it should return 0 if the user gets a question wrong, and 1 if the user gets a question right. Then `numSuccesses` just tallies the number of correct answers and exits the `While` loop when three correct answers have been given. `PracticeDifferentiation` is defined as follows:

```
PracticeDifferentiation[ ]:=
    Module[{prob, probAns, userAns},
        {prob, probAns} =
            GenerateRandomProblemAndAnswer[ ];
        userAns          = AskUser[prob];
        If[IsAnswerOkQ[userAns, probAns],
           (Print["Looks right to me!"];
            1
        ),

           (Print["That's not what I got."];
            Print["You said the derivative of ",
                prob, " wrt x is ", userAns];
            Print["The derivative of ", prob,
               " wrt x is ", probAns];
            0
        )
        ]
    ]
```

The meat of the program consists of four distinct phases. *Mathematica* generates a random differentiation problem, poses the problem to the user, obtains the reply, and checks whether that reply is correct. The problem-generation phase uses a set of functional forms. One form is picked at random; an instance of that form is created and then differentiated. These last three steps therefore generate both a problem and its correct answer. The code follows:

```
GenerateRandomProblemAndAnswer[ ]:=
   Module[{expr, answer},
        expr   = GenerateExpression[];
        answer = Simplify[Expand[D[expr, x]],
                     Trig->True];
        {expr, answer}
   ]

GenerateExpression[ ]:=
    Module[{fns, form, instance},
        fns  = {Sin, Cos, Tan, Log, Exp, Sqrt};
        form = RandomElement[{f[x], f[x]+g[x],
                  f[x]-g[x], f[x]^x, x^f[x],
                  f[x]^g[x], f[x]/g[x], f[g[x]]}];
        instance = form /.
            {f -> RandomElement[fns],
             g -> RandomElement[fns]};
        Simplify[Expand[instance], Trig->True]
   ]

RandomElement[list_List]:=
    list[[ Random[Integer, {1,Length[list]}] ]]
```

The call to `Random` generates a random integer between 1 and the number of functional forms inclusive. The program then uses this number to obtain a random element of the list of functional forms (the `list[[`*i*`]]` command returns the ith element of the list). In other programs, you may want to call `Random` to produce different types of random numbers. The general form is `Random[`*type*`,` *range* `]`, where *type* can be `Integer`, `Real`, or `Complex`. By default, `Random` picks numbers uniformly in the given range. If you want to pick numbers according to some other distribution, then you must first load one of the *Mathematica* packages `` Statistics`ContinuousDistributions` `` or `` Statistics`DiscreteDistributions` ``, depending on the kind of random number you desire. Thereafter, you can call `Random` with a distribution as its argument. For example, `Random[NormalDistribution[mu, sigma]]` would generate a random real number according to a normal distribution, with mean `mu` and standard deviation `sigma`.

Once the functional form is selected, it is instantiated by replacing the token function symbols `f` and `g` with actual function names such as `Sin`, `Cos`, and so on.

The expression is then differentiated and the result tidied up, using the *Mathematica* command `Simplify`, before returning a problem/answer pair.

Now we are ready to pose the question to the user and to obtain the reply. The trick here is to call `Input` with a string as argument that contains the text of our question.

We use the notation `<>`, a shorthand notation for the command `StringJoin`, to construct a string of the question that will be posed to the user.

```
AskUser[problem_]:=
    Input["What is the derivative of " <>
          ToString[InputForm[problem]] <>
          " with respect to x?"
    ]

IsAnswerOkQ[userAns_, probAns_] :=
    Simplify[Expand[userAns],Trig->True]===probAns
```

Run this program by typing `TakeTest[ ]`. One problem with creating interactive programs using `Input` is that the response must employ correct *Mathematica* syntax. If the user types an ill-formed expression, *Mathematica* will complain. In certain applications you may want to allow users to type incorrect expressions, as in, for example, a program designed to test whether people know *Mathematica* syntax. In this case, you should use `InputString` in conjunction with a parser that you will have to write. You will find the *Mathematica* command `ReadList` useful for parsing strings.

13.3 Creation of Animated Graphics in *Mathematica*

Example 13.5 In addition to programming mathematical operations in *Mathematica*, you can also program graphics operations. We have already seen how you can create graphs of functions; now we will show you how to create animations that illustrate important concepts in calculus. The following program creates an animation of the convergence of Taylor polynomial approximations to a function.

To begin, we generate a sequence of Taylor approximations to the function $\sin(x) + \cos(2x)$, centered at the $x = 0$ from first order to seventh order. The *Mathematica* command for generating a Taylor series is `Series`. This command creates a `SeriesData` object that contains both the polynomial approximation to the desired order and a term that indicates the order of the remainder. We use `Normal` to convert the output from `Series` into a regular polynomial approximation.

```
In[14]:=  ColumnForm[
              Table[
                  Normal[Series[Sin[x]+Cos[2 x],{x,0,i}]],
                  {i,1,7}
              ]
          ]

Out[14]=  1 + x
                  2
          1 + x - 2 x
                         3
                  2     x
          1 + x - 2 x  - --
                         6
                         3      4
                  2     x    2 x
          1 + x - 2 x  - -- + ----
                         6     3
                         3      4     5
                  2     x    2 x     x
          1 + x - 2 x  - -- + ---- + ---
                         6     3     120
                         3      4     5       6
                  2     x    2 x     x     4 x
          1 + x - 2 x  - -- + ---- + --- - ----
                         6     3     120    45
                         3      4     5       6       7
                  2     x    2 x     x     4 x       x
          1 + x - 2 x  - -- + ---- + --- - ---- - ----
                         6     3     120    45    5040
```

To create the animation, we first construct its frames. In each frame, we plot both the true function (shown as a dashed line) and the truncated Taylor series approximation to that function. As we increase the order of the Taylor series approximation, we see that the truncated series becomes a closer approximation to the function.

```
AnimateTaylorApproximations[fn_, x_, x0_, n_,
        plotRange_]:=
    Module[{taylorPolys},
        taylorPolys = Table[
            Normal[Series[fn, {x,x0,i}]],
            {i,1,n}];
        Map[Plot[{#, fn},
            {x, plotRange[[1,1]], plotRange[[1,2]]},
                    PlotRange->plotRange,
                    PlotStyle->
                {GrayLevel[0],Dashing[{0.02,0.02}]}
            ]&, taylorPolys
        ]
    ]
```

Try giving the command

```
AnimateTaylorApproximations[Sin[x] + Cos[2x], x,0,15,
    {{-6,6}, {-10,10}}]
```

If you are using the Notebook Front End interface, once all the frames have been computed, double-clicking on any one frame causes all the frames to be displayed in rapid succession. You will see that the higher-order Taylor series approximations agree with the true function over a progressively wider range.

13.4 Concluding Remarks

We have given only a brief glimpse of the programming capabilities of *Mathematica*. You should explore *Mathematica*'s ability to read data from files, to run and manipulate programs outside of *Mathematica*, and to export portions of your *Mathematica* sessions in a variety of useful formats. Finally, you might have one important question that we have not addressed: namely, if you create a program in *Mathematica*, then you might naturally ask how you can use it in a different *Mathematica* session. The answer is simple. Take the *Mathematica* programs that you create and place them in one or several files. For example, you might place the *Mathematica* program `AnimateTaylorApproximations` with any other procedures related to Taylor polynomials in a file named `taylor.m`. When you want to use these programs, simply invoke the *Mathematica* command `<<taylor.m`. Every *Mathematica* function in that file will be read into your *Mathematica* session and will be available for your use. When you save your program definitions in a *Mathematica* Notebook, we recommend converting the cells that contain definitions to initialization cells. Then, when you read the contents of your Notebook named *notebook*`.ma` into your *Mathematica* session using `<<` *notebook*`.ma`, or when you open the Notebook, the definitions will be automatically loaded.

14 Troubleshooting Tips

You have now seen the essentials of programming in *Mathematica*. You should therefore be able to use *Mathematica* interactively and to write programs in *Mathematica*. What do you do when *Mathematica* does not give you results that you expect? How do you figure out what went wrong? In this chapter we discuss common traps and pitfalls that catch *Mathematica* users. We also describe tools for debugging.

14.1 The Top 10 Traps for Novice *Mathematica* Users

Mathematica has its own idiosyncrasies, and certain problems seem to arise, repeatedly, for just about everybody. We hope that, by pointing out common errors to you, along with examples of how they occur, how to spot them, and how to avoid them, we will help you achieve proficiency with this powerful software tool quickly and painlessly. Here is a list of traps that catch many new *Mathematica* users.

1. Missing or incorrect punctuation.
2. Missing or incorrect names or arguments.
3. Referencing previous results.
4. Specifying input that is longer than the width of the screen.
5. Confusing exact and approximate calculations.
6. Having trouble using on-line help.
7. Forgetting to save your results.
8. Forgetting to load packages.
9. Trying to get *Mathematica* to do too much.
10. Forgetting to check results for plausibility.

Before considering specific examples of common errors, be aware that you can prevent many errors simply by increasing the font size at the beginning of a *Mathematica* session. This technique sounds trivial, but it is amazing how many students do not think about it, and suffer eyestrain and wasted time correcting errors caused by reading difficulties. In the same vein, some people have found it helpful to insert blank spaces generously in lines of *Mathematica* statements to improve readability.

Now we will show you some errors that have caused problems to us, our colleagues, and our students.

14.2 Missing or Incorrect Punctuation

Missing or incorrect punctuation can give you surprising results.

Omitting a Comma

Look at what happens when you omit a comma in an `If` statement. You might think that the function `f` would return the absolute value of its argument.

```
In[1]:=  Clear[f]
         f[x_] := If[x > 0, x  -x]
```

In fact, this function returns `x - x` or 0 if `x` > 0 and returns `Null` if `x` ≤ 0, because a comma is missing between two arguments in the `If` statement.

```
In[3]:=  f[3]

Out[3]=  0

In[4]:=  f[-3]
```

Forgetting a Space

Many users have made the mistake of using a symbol consisting of two characters to represent a product. The notation `xy` does not represent the product of `x` and `y`; it represents a symbol whose name consists of two characters.

Using = Instead of ==

Suppose that you want to solve the equation $y = mx + b$ for x, and you enter

```
In[5]:=  Clear[a, m, x, y]
         Solve[y = m x  + b, x]

Solve::eqf: b + m x is not a well-formed equation.

Out[6]=  Solve[b + m x, x]
```

then the expression `m x + b` is assigned the name `y`.

Now, if we enter the equation properly (with two equal signs),

```
In[7]:=  Solve[y == m x  + b, x]

Out[7]=  {{}}
```

Mathematica does not return any solutions. Why not? Because `y` is replaced by the expression `m x + b`. Then, *Mathematica* attempts to solve the equation $mx + b = mx + b$, which simplifies to an equation with no variables: $0 = 0$. So, no solutions are found. After we clear the value of `y`, `Solve` returns a solution.

```
In[8]:=  Clear[y]
         Solve[y == m x  + b, x]

Out[9]=          -b + y
         {{x -> ------}}
                   m
```

Missing Parentheses

Mathematica makes unmatched parentheses fairly easy to find, as the following example shows:

```
In[10]:=  x Sqrt[2x + 1] / ((x^2 + 5)(x + 1)

Syntax::sntxi: Incomplete expression.
```

The following example shows how you can use parentheses to ensure that certain calculations are carried out as intended. Suppose that you want *Mathematica* to calculate $\sum_{k=1}^{10} \frac{1}{k(k+2)}$, and you enter

```
In[11]:=  Sum[1 / k (k+2), {k, 1, 10}]

Out[11]=  19981
          -----
          1260
```

This answer is not the value of $\sum_{k=1}^{10} \frac{1}{k(k+2)}$. Instead, it is the value of $\sum_{k=1}^{10} \frac{(k+2)}{k}$. Remember that two expressions separated by a space are multiplied together. Using parentheses, we instruct *Mathematica* to divide by $k + 2$.

```
In[12]:=  Sum[1/(k(k+2) ), {k, 1, 10}]

Out[12]=  175
          ---
          264

In[13]:=  N[%]

Out[13]=  0.662879
```

Using Parentheses Instead of Brackets or Vice Versa

Using parentheses, `( )`, in place of brackets, `[ ]`, can give substantially different results. The notation `foo[0]` calls the function `foo` with `0` as its argument. The notation `foo(0)` gives the product of `0` and the symbol `foo`.

```
In[14]:=  foo[0]

Out[14]=  foo[0]

In[15]:=  foo(0)

Out[15]=  0
```

Let's look at how *Mathematica* represents the expression `foo(a)`, to understand why `foo(0)` returns 0.

```
In[16]:=  foo(a)

Out[16]=  a foo
```

The expression `foo(a)` is taken as the product of `a` and `foo`.

14.3 Missing or Incorrect Names or Arguments

There are several ways in which a function might not behave as you would expect. If the function is returned unevaluated, *Mathematica* either was unable to evaluate the function or does not know how to evaluate it. Make sure that you call the function correctly by using the number and type of arguments expected. Be sure to spell the function name correctly. You can check the name of the function and number of arguments with the built-in help facility. For example, when you are integrating or differentiating, do not forget to specify the variable of integration or differentiation.

```
In[17]:=  D[x^2]

Out[17]=   2
          x
```

Why did *Mathematica* return x^2? The function `D` differentiates the first argument with respect to the second argument, third argument, and so on. For example, `D[x^2 y^3, x, y]` returns $\frac{d}{dy}\frac{d}{dx}x^2y^3$. The expression `D[x^2]` instructs *Mathematica* to differentiate x^2 zero times. *Mathematica* simply returns the expression x^2 itself.

14.4 Referencing Previous Results

You can refer to a previous result in a given *Mathematica* session by using one or more percent signs; that is, % refers to the last result, %% refers to the second to last result, %%% refers to the third to last result, and so forth.

When using the Notebook Front End, you can execute inputs in whatever order you desire. So, the result at the top of your screen may have been executed after the result at the bottom of your screen. When referencing a previous result, you should realize that a percent sign refers to the last result returned by the *Mathematica* kernel (the computational engine), which may not be the result that appears immediately above your current input.

Reexecuting Statements to Avoid Retyping Them

Suppose that you want to calculate the areas of two circles of radii 2 and 3 respectively. You enter

```
In[18]:=  r = 2;

In[19]:=  A = Pi r^2;

Out[19]=  4 Pi

In[20]:=  N[%]

Out[20]-  12.5664

In[21]:=  r = 3;
```

If you now reexecute the `N` command, you will not get the correct answer.

```
In[22]:=  N[%]

Out[22]=  3.
```

The last % refers to the last output, which was 3, rather than 9π. You might try entering N[A], but this is not correct either, since A has not been recalculated. The remedy is first to reexecute the statement where A is calculated, and then to reexecute the N command.
Hint: Putting several statements on one line or in a cell ensures that they will all be reexecuted when the line is reentered.

```
In[23]:=  r = 2;    A = Pi r^2;
          N[%]

Out[24]=  12.5664
```

Now, when r = 2 is changed to r = 3 and the line is reexecuted, the N[%] command will work as you want it to. Misusing % is a common mistake, which you can be avoid by assigning names to your results and referring to those names, rather than using %.
Hint: Another way to avoid retyping is to use Copy and Paste from the Edit menu or to type type <COMMAND> –l (the command key and the lowercase letter L) with your cursor in a cell just below the input cell that you want to copy.

As another example of reexecuting a statement, suppose that you have entered the statement

```
In[25]:=  Integrate[x^2, x]

Out[25]=   3
          x
          --
          3
```

A variable assigned a value retains that value until it is explicitly reassigned or cleared. Here, we assign the value 1 to x.

```
In[26]:=  x = 1

Out[26]=  1
```

When we integrate an expression with respect to x, *Mathematica* prints an error message, because the variable x is replaced with its value. In essence, we are asking *Mathematica* to integrate a number with respect to another number. That makes no sense.

```
In[27]:=  Integrate[x^2, x]

Integrate::ilim:
   Integration limit 1 is not of the form {x,xmin,xmax}.

Out[27]=  Integrate[1, 1]
```

To avoid using predefined variables, clear the value that you assign to a variable as soon as you finish with it or just before you use it, and, whenever possible, declare variables to be local to a function by using `Module` or `Block`.

```
In[28]:=  Clear[x]
          Integrate[x^2, x]

Out[29]=   3
          x
          --
          3
```

14.5 Specifying Input That is Longer Than the Width of the Screen

If a command can be considered complete before a carriage return, *Mathematica* evaluates the line even if the expression continues on the next line. Notice that the following input returns 6, rather than 15 as you might expect.

```
In[30]:=  a = 4
              + 5
              + 6

Out[32]=  6
```

What is going on? In the first line, *Mathematica* assigns 4 to a. Next, *Mathematica* returns a 5; then it returns a 6. Because the 6 is the last result calculated, it is printed. As you can see, a is assigned the value 4.

```
In[33]:=  a

Out[33]=  4
```

If *Mathematica* cannot parse the input after reading a line, it reads the next line of input. When a line ends with a plus sign, +, or some other character that indicates there is more input to be parsed, *Mathematica* reads the next input line.

```
In[34]:=  a = 4 +
              5 +
              6

Out[34]=  15
```

Now a is assigned the value 15.

```
In[35]:= a

Out[35]= 15
```

14.6 Confusing Exact and Approximate Calculations

Many commands work differently depending on whether *Mathematica* is computing in exact arithmetic. A decimal point in a single number is enough to cause *Mathematica* to use approximate arithmetic. Consider the following examples.

Some commands, such as `FactorInteger` and `Limit`, do not work correctly on approximate arguments.

```
In[36]:= FactorInteger[124]

Out[36]= (-1 + x) (1 + x)

In[37]:= FactorInteger[124.]

FactorInteger::facn:
   First argument 124. in FactorInteger[124.]
     is not an exact number.

Out[37]= FactorInteger[124.]
```

`Limit` returns the correct answer when the input is exact.

```
In[38]:= Limit[(1 + 1/x)^x, x -> Infinity]

Out[38]= E
```

Notice that `Limit` returns the wrong answer when the input contains an approximate number.

```
In[39]:= Limit[(1. + 1/x)^x, x -> Infinity]

Out[39]= 0
```

Polynomials of degree four or less can be solved exactly

```
In[40]:=  Solve[x^4 + x - 2 == 0, x]

Out[40]=  {{x -> 1},
                                    1/3
                 1               2 2
           {x-> -(-) - ------------------------ +
                 3                          1/3
                         3 (-47 + 3 Sqrt[249])

                                        1/3
                   (-47 + 3 Sqrt[249])
                   ----------------------
                                 1/3
                              3 2
           },
                          1/3
                  1      2     (1 + I Sqrt[3])
           {x -> -(-) + ------------------------ -
                  3                          1/3
                          3 (-47 + 3 Sqrt[249])

                                                        1/3
               (1 - I Sqrt[3]) (-47 + 3 Sqrt[249])
               ---------------------------------------
                                     1/3
                                  6 2
           },
                          1/3
                  1      2     (1 - I Sqrt[3])
           {x -> -(-) + ------------------------ -
                  3                          1/3
                          3 (-47 + 3 Sqrt[249])

                                                        1/3
               (1 + I Sqrt[3]) (-47 + 3 Sqrt[249])
               ---------------------------------------
                                     1/3
                                  6 2
           }
          }
```

They can also be solved approximately:

```
In[41]:=  Solve[x^4. + x - 2 == 0, x]

Out[41]=  {{x -> -1.35321}, {x -> 0.176605 - 1.20282 I},
            {x -> 0.176605 + 1.20282 I}, {x -> 1.}}
```

On the other hand, polynomials of degree five or higher cannot be solved exactly in every case. Here's how *Mathematica* responds when asked for exact solutions to such equations:

```
In[42]:=  Solve[x^5 - x == 0, x]   (* easy to solve since *)
                                   (* it can be factored  *)

Out[42]=  {{x -> -1},{x -> 0},{x -> -I},{x -> I},{x -> 1}}

In[43]:=  Solve[x^5 + x - 3 == 0, x](* not easy to solve *)

Out[43]=                         5
          {ToRules[Roots[x + x  == 3, x]]}
```

However, *Mathematica* has a variety of ways for finding approximate roots. Recall that a fifth-degree polynomial has five roots, counting multiplicities, and that the complex roots of a real polynomial come in complex-conjugate pairs.

```
In[44]:=  N[%]

Out[44]=  {{x -> -1.04188 - 0.82287 I},
            {x -> -1.04188 + 0.82287 I},
            {x -> 0.475381 - 1.1297 I},
            {x -> 0.475381 + 1.1297 I}, {x -> 1.133}}
```

The following command also returns approximations to the solutions (notice the decimal point in the 3.0).

```
In[45]:=  Solve[x^5 + x - 3.0 == 0, x]

Out[45]=  {{x -> -1.04188 - 0.82287 I},
            {x -> -1.04188 + 0.82287 I},
            {x -> 0.475381 - 1.1297 I},
            {x -> 0.475381 + 1.1297 I}, {x -> 1.133}}
```

`NSolve` also returns all the roots.

```
In[46]:=  NSolve[x^5 + x - 3 == 0, x]

Out[46]=  {{x -> -1.04188 - 0.82287 I},
            {x -> -1.04188 + 0.82287 I},
            {x -> 0.475381 - 1.1297 I},
            {x -> 0.475381 + 1.1297 I}, {x -> 1.133}}
```

Here is another example that illustrates a difference between results calculated using exact and approximate inputs.

```
In[47]:=  Solve[Exp[x] + x - 2 == 0, x]

Solve::tdep:
   The equations appear to involve transcendental
     functions of the variables in an essentially
     non-algebraic way.

Out[47]=                x
          Solve[-2 + E  + x == 0, x]

In[48]:=  FindRoot[Exp[x] + x - 2. == 0, {x, 1}]

Out[48]=  {x -> 0.442854}
```

14.7 Having Trouble Using On-Line Help

New users sometimes complain about *Mathematica*'s on-line help. Other new users do not bother to use the on-line help. This is unfortunate, because the on-line help is a valuable resource, whether used as reference for looking up forgotten syntax, or as a vehicle for exploring the myriad commands and mathematics *Mathematica* puts at your disposal. We shall look at a few typical situations.

Suppose that you want to plot two graphs on one coordinate system, but you do not remember the exact syntax. To learn about the `Plot` command, enter

```
In[49]:=  ?Plot

Plot[f, {x, xmin, xmax}] generates a plot of f as a
   function of x from xmin to xmax. Plot[{f1, f2, ...},
   {x, xmin, xmax}] plots several functions fi.
```

Notice that the usage statement includes two templates for `Plot`. The second template shows that the first argument can be a list of functions or expressions.

```
In[50]:=  Plot[{Sin[x], Cos[x]}, {x, 0, 2Pi}];
```

If you want to see the properties or attributes and the options for a function, enter

```
In[51]:=  ??Plot

Plot[f, {x, xmin, xmax}] generates a plot of f as a
   function of x from xmin to xmax. Plot[{f1, f2, ...},
   {x, xmin, xmax}] plots several functions fi.

Attributes[Plot] = {HoldAll, Protected}

Options[Plot] =
  {AspectRatio -> GoldenRatio^(-1), Axes -> Automatic,
   AxesLabel -> None, AxesOrigin -> Automatic,
   AxesStyle -> Automatic, Background -> Automatic,
   ColorOutput -> Automatic, Compiled -> True,
   DefaultColor -> Automatic, Epilog -> {},
   Frame -> False, FrameLabel -> None,
   FrameStyle -> Automatic, FrameTicks -> Automatic,
   GridLines -> None, MaxBend -> 10.,
   PlotDivision -> 20., PlotLabel -> None,
   PlotPoints -> 25, PlotRange -> Automatic,
   PlotRegion -> Automatic, PlotStyle -> Automatic,
   Prolog -> {}, RotateLabel -> True, Ticks -> Automatic,
   DefaultFont :> $DefaultFont,
   DisplayFunction :> $DisplayFunction}
```

Sometimes, the explanations contain terms that may be unfamiliar to you. It may be best to ignore what you do not understand; you will either learn it later, or else find that you do not need to know it. For example, suppose that you want to find out about the `Map` command. You enter

```
In[52]:=  ?Map

Map[f, expr] or f /@ expr applies f to each element on the
   first level in expr. Map[f, expr, levelspec] applies f
   to parts of expr specified by levelspec.
```

You may not understand what is meant by *first level.* You might try using the function.

```
In[53]:=  Map[f, a + b + c]

Out[53]=  f[a] + f[b] + f[c]
```

Mathematica's on-line help does not attempt to teach mathematics, so you have to learn to separate the mathematics topics from the *Mathematica* topics. For example, the entry

```
In[54]:=  ?Integrate
```

will tell you about *Mathematica*'s `Integrate` command for evaluating *definite integrals*, *indefinite integrals*, and *multiple integrals*, but it expects you already to know about these types of integrals.

Sometimes unexpected functions comes up, but they can safely be ignored, as in the following example. You want to evaluate the integral $\int \cos(x)/x\, dx$, so you enter the following:

```
In[55]:=  Integrate[Cos[x]/x, x]

Out[55]=  CosIntegral[x]
```

"What in the world is `CosIntegral`?" you say to yourself. To find out, you decide to enter

```
In[56]:=  ?CosIntegral
```

You get a bit of information.

Mathematica's on-line help is like the on-line help for Unix in that, if you know the name of the command that you want, then you can find more information about it. However, if you do not know what command you need, you may have trouble finding its name.

Hint: The Help Browser that is available on some versions of *Mathematica* lets you locate the particular help entry without needing to know the name. All the commands that are in the Help Browser are arranged categorically. So, if you know the type of command that you need, you can browse the commands and have a chance at finding the one that you need. The *Mathematica* Help Stack, published by Variable Symbols, Inc., is also a great source of information about commands. The Help Stack contains examples of all the commands built into *Mathematica*.

Incomprehensible Error Messages

Mathematica prints error messages when your input does not match its own internal rules. Error messages are of the form

Symbol::errorTag: A brief message describing the error.

For example, when we use the assignment statement, =, instead of the ==, for testing whether $1 + 1$ is equal to 3, *Mathematica* prints the following messages:

```
In[57]:=  1 + 1 = 3

Set::write: Tag Plus in 1 + 1 is Protected.
```

Wolfram Research, the developer of *Mathematica*, publishes the guide *Mathematica Warning Messages*, which contains a listing of most of *Mathematica*'s error messages. The guide also includes a brief explanation of situations in which the message might appear, examples of input that produce the message, and suggestions for dealing with the problem.

When *Mathematica* can interpret your input, it will do so even when your instructions do not do what you intended.

Dividing by Zero

Mathematica warns you when you divide an expression by zero, even when the expression has a definite value.

```
In[58]:=  Plot[Sin[x]/x, {x, -15, 15}];

                                         1
Power::infy: Infinite expression -- encountered.
                                         0.
Infinity::indet:
   Indeterminate expression 0. ComplexInfinity
     encountered.
Plot::plnr:
   CompiledFunction[{x}, <<1>>, -C<<9>>de-][x]
     is not a machine-size real number at x = 0..
```

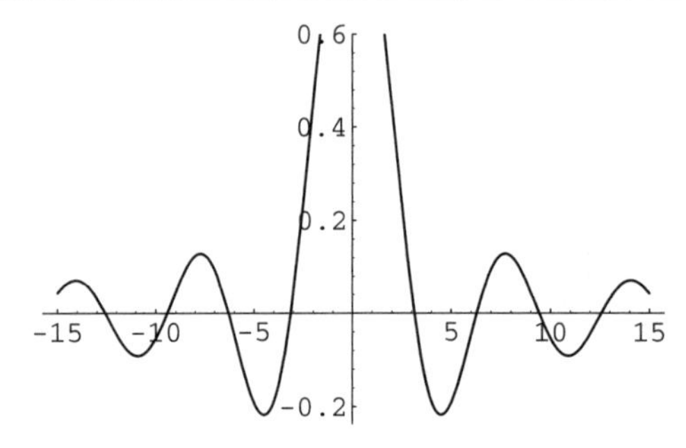


```
In[59]:=  Limit[Sin[x]/x, x -> 0]

Out[59]=  1
```

Spelling Errors

The spelling checker warns a user when a new symbol is spelled in a similar way to an existing symbol. For example, here we invoke the function `sin[2Pi]`, which is different from the built-in `Sin` function.

```
In[60]:=  sin[2Pi]

General::spell1:
   Possible spelling error: new symbol name "sin" is similar
      to existing symbol "Sin".

Out[60]=  sin[2 Pi]
```

```
In[61]:=  Sin[2Pi]

Out[61]=  0
```

If you find the spelling checker annoying, you can turn it off by typing

```
In[62]:=  Off[General::spell1, General::spell]
```

Doing Things in the Wrong Order

The order in which *Mathematica* evaluates arguments can affect the result. Suppose that you want to define a function that computes the derivative of the expression $x \cos x$ with respect to x, and then substitutes the argument into the resulting expression. Here is how you might define such a function.

```
In[63]:=  f[x_] := D[x Cos[x], x]
```

Note that when you call `f[2]`, it does not return the value that you would expect, because *Mathematica* replaces all occurrences of `x` with the value of the argument, (e.g., 2), and then tries to take the derivative of the resulting expression.

```
In[64]:=  f[2]

General::ivar: 2 is not a valid variable.

Out[64]=  D[2 Cos[2], 2]
```

Using `Evaluate`, you can force *Mathematica* to evaluate the derivative before replacing `x` with the value of the argument.

```
In[65]:=  g[x_] := Evaluate[D[x Cos[x], x]]
          g[2]

Out[66]=  Cos[2] - 2 Sin[2]
```

Here, we attempt to graph $\sin x$, together with its derivative.

```
In[67]:= Plot[{Sin[x], D[Sin[x], x]}, {x, 0, 2Pi}];

General::ivar: 0. is not a valid variable.
General::ivar: 0. is not a valid variable.
Plot::plnr: CompiledFunction[{x}, <<1>>, -CompiledCode-][x]
     is not a machine-size real number at x = 0..
General::ivar: 0.261799 is not a valid variable.
General::stop:
   Further output of General::ivar
     will be suppressed during this calculation.
Plot::plnr: CompiledFunction[{x}, <<1>>, -CompiledCode-][x]
     is not a machine-size real number at x = 0.261799.
Plot::plnr: CompiledFunction[{x}, <<1>>, -CompiledCode-][x]
     is not a machine-size real number at x = 0.523599.
General::stop:
   Further output of Plot::plnr
     will be suppressed during this calculation.
```

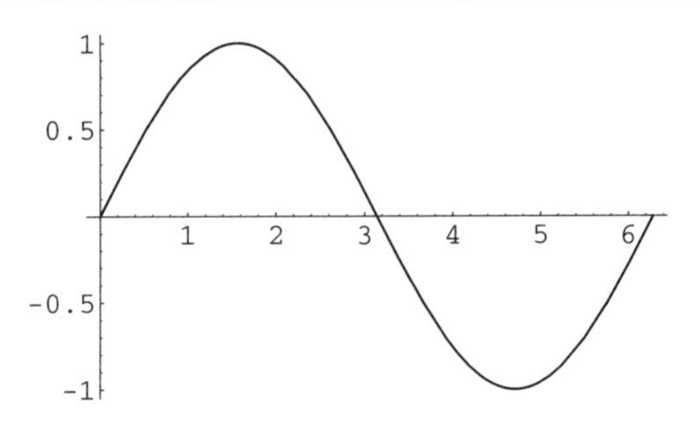

Note that *Mathematica* plots $\sin x$ without a problem, but it does not plot $\sin x$'s derivative. As in the previous example, *Mathematica* substitutes a value into the expression `D[Sin[x], x]` before computing the derivative. Once again, we use `Evaluate` to force *Mathematica* to evaluate the expression before plugging in values to obtain points along the curve.

```
In[68]:= Plot[Evaluate[{Sin[x], D[Sin[x],x]}], {x,0,2Pi}];
```

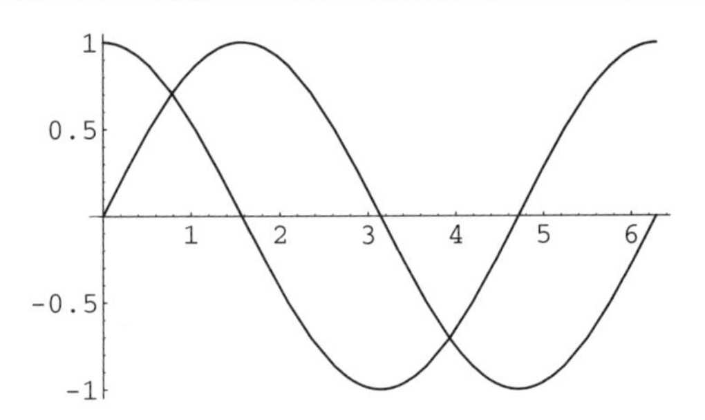

14.8 Forgetting to Save Your Results

It sometimes happens that you want to interrupt a *Mathematica* session for one reason or another. For example, a computation may take longer than you anticipated. If you do not save the session, you may lose work and be forced to start from scratch. For this reason, it is good practice to open the `File` menu and select `Save`, to save your work, regularly. For example, suppose that you have been working for some time and have generated numerous lines of *Mathematica* output. Then, you want to calculate $\sum_{n=0}^{200} n!$, but you mistakenly enter

```
In[69]:=  Sum[n!, {n, 0, 20000}]
```

This calculation will probably take longer than you want to wait, so you might abort the calculation and exit, without saving the session. Then, you would lose all the results you computed since you last saved the session.

14.9 Forgetting to Load Packages

Some *Mathematica* commands are defined in packages that must be loaded into a *Mathematica* session before the commands can be used. To see a listing of the available packages, refer to the *Mathematica Quick Reference* (Variable Symbols and Addison-Wesley) or to the *Guide to Standard Mathematica Packages* that comes with your copy of *Mathematica*.

Sometimes, you may remember a command and its syntax, but forget that it is part of a package. If you try to use the command before loading the package, not only will it not work, but the reason may also not be readily apparent to you. For example, suppose that you want to plot several sets of data points, along with the graph of a function given by an expression, and you remember that the `MultipleListPlot` is designed for such a purpose.

```
In[70]:=  MultipleListPlot[{1,2,4,8}, {2,4,6,8},
               PlotJoined -> True]

Out[70]=  Out[1]=
          MultipleListPlot[{1, 2, 4, 8}, {2, 4, 6, 8},
            PlotJoined -> True]
```

Note that *Mathematica* does not know anything about this function, so it creates the symbol `MultipleListPlot` and returns exactly what you entered. The result was not what you wanted. Now, let's load the package `Graphics`MultipleListPlot``.

```
In[71]:=  Needs["Graphics`MultipleListPlot`"]

MultipleListPlot::shdw:
  Warning: Symbol MultipleListPlot
    appears in multiple contexts
   {Graphics`MultipleListPlot`, Global`}; definitions in
    context Graphics`MultipleListPlot`
    may shadow or be shadowed by other definitions.
```

Note that *Mathematica* warns us that there are multiple definitions of the symbol and that one might shadow the other. In other words, you might call one definition instead of the other. When we ask about `MultipleListPlot`, we are told about the symbol that *Mathematica* created recently, which is not the one that we want.

```
In[72]:=  ?MultipleListPlot

Global`MultipleListPlot
```

Why does *Mathematica* not show us the usage statement? Because the symbol `MultipleListPlot` that we are calling does not have a usage statement. The definition in the package `Graphics`MultipleListPlot`` has a usage statement, as you can see.

```
In[73]:=  ?Graphics`MultipleListPlot`MultipleListPlot

MultipleListPlot[ l1, l2, ... ] allows many lists of data
   to be plotted on the same graph.  Each list can be
   either a list of pairs of numbers, in which case the
   pairs are taken as x,y-coordinates, or else a list of
   numbers, in which case the numbers are taken as
   y-coordinates and successive integers starting with 1
   are supplied as x-coordinates.  The DotShapes option
   specifies plotting symbols to be used for the lists of
   data, and (if the option PlotJoined->True is specified)
   the option LineStyles specifies styles for the lines
   connecting the points.
```

If we remove, or delete, the symbol that we created earlier, then we can access the definition that we want.

```
In[74]:=  Remove[MultipleListPlot]
          ?MultipleListPlot

MultipleListPlot[ l1, l2, ... ] allows many lists of data
   to be plotted on the same graph.  Each list can be
   either a list of pairs of numbers, in which case the
   pairs are taken as x,y-coordinates, or else a list of
   numbers, in which case the numbers are taken as
   y-coordinates and successive integers starting with 1
   are supplied as x-coordinates.  The DotShapes option
   specifies plotting symbols to be used for the lists of
   data, and (if the option PlotJoined->True is specified)
   the option LineStyles specifies styles for the lines
   connecting the points.

In[76]:=  MultipleListPlot[{1,2,4,8}, {2,4,6,8},
             PlotJoined -> True];
```

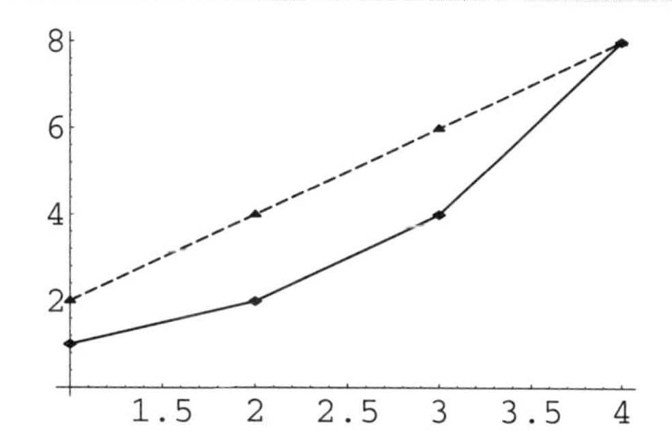

14.10 Trying to Get *Mathematica* to Do Too Much

When students are given an assignment to use *Mathematica* on a problem, they often proceed under the assumption that they must use *Mathematica* for every step in the problem. This assumption can lead to the absurd situation where they painstakingly cajole *Mathematica* into performing steps that are completely obvious, and should just be done by hand. As an illustration of this point, we consider the following example. Suppose that we are told that the temperature $T = T(t)$ of a cooling cup of coffee, where t denotes time, satisfies the equation $\ln(T - 70.) = -k\,t + c$, where k and C are unknown constants. We are also told that the temperature at $t = 0$ is 210, and that $T(1) = 190$. We want to solve for $T(t)$, and also for the time $t = t_1$ when $T(t_1) = 150$. The following statements show one way to use *Mathematica* to solve this problem, but most beginning *Mathematica* users would probably choose to do some of the steps themselves—for instance, making the observation that $c = \ln(210. - 70.) = \ln(140)$)—rather than finding the commands to make *Mathematica* do them all.

```
In[77]:=  eq = Log[T - 70.] == -k t + c

Out[77]=  Log[-70. + T] == c - k t

In[78]:=  parameters = Solve[{eq /. {T -> 210, t -> 0},
                       eq /. {T -> 190, t -> 1}}, {c, k}]

Out[78]=  {{k -> 0.154151, c -> 4.94164}}
```

Note: Although it appears that we have solved for k and c, the following shows that the numerical values have not yet been assigned to k and c:

```
In[79]:=  {k, c}

Out[79]=  {k, c}
```

The following commands assign the values to k and c, and then complete the problem.

```
In[80]:=  {{k, c}} = {k, c} /. parameters

Out[80]=  {{0.154151, 4.94164}}


In[81]:=  {T} = T /. Solve[eq, T]

Out[81]=                        4.94164 - 0.154151 t
          {70. + 1. 2.71828                          }


In[82]:=  t1 =  {t} /. FindRoot[T == 150., {t, 1}]

Out[82]=  {3.63032}
```

14.11 Forgetting to Check Results for Plausibility

When your *Mathematica* code does not give the result you expect, answering the following questions might prove helpful in debugging your input.

- Did you type a lowercase letter where you should have typed an uppercase letter, or vice versa?
- Does each bracket, parenthesis, and brace have a mate?
- Have any variables been assigned values inadvertently?
- Did you call the function that you intended?

- Does each statement work as intended?
- Did you call functions with the correct number and type of arguments?
- Have you forgotten any punctuation?
- Are there semicolons or commas between statements or elements?
- Does the code work on test data?
- Have you loaded all relevant packages?

In addition to the preceding checklist, there are also a few functions that can be used for debugging *Mathematica* programs.

Command	**Description**
`On, Off`	Tracing
`Print`	A primitive debugging tool
`Trace`	Generate a list of intermediate expressions
`Input`	In conjunction with `?`

Examples of each of these functions are provided next.

Tracing

When you are interested in seeing the steps *Mathematica* takes to arrive at a result, use the tracing capability.

Let's define a couple of functions.

```
In[83]:=  Clear[f, g]
          f[x_] := x^2 + g[x]
          g[x_] := x + 9
```

Now let us turn tracing on.

```
In[86]:=  On[f, g]
```

You can see the steps that *Mathematica* takes to arrive at a result for `f[3]`.

```
In[87]:=  f[3]

                        2
f::trace: f[3] --> 3  + g[3].
g::trace: g[3] --> 3 + 9.

Out[87]=  21
```

Turn tracing off with `Off`.

```
In[88]:= Off[f, g]
```

Print

The `Print` function allows you to look at intermediate results.

```
In[89]:= median[list_List] :=
              Module[{
                         sl,
                         len
                     },
                     len = Length[list];
                     Print["Length of list: ", len];
                     sl = Sort[list];
                     Print["Sorted list: ", sl];
                     If[
                         OddQ[
                             Length[sl]
                         ],
                         sl[[ (len + 1)/2 ]],
                         (sl[[len/2]] + sl[[len/2+1]])/2
                     ]
              ]

Out[89]= median[{43, 45, 23, 65}]

Out[89]= Length of list: 4
         Sorted list: {23, 43, 45, 65}

Out[89]= 44
```

Trace

The command `Trace` returns a list of the intermediate expressions computed when evaluating its argument. The output allows you to see the steps that *Mathematica* takes to arrive at a result. Note that *Mathematica* first evaluates the product of 2 and 3, then evaluates the product of 5 and 7, and then adds the two values to arrive at the result: 41.

```
In[90]:= Trace[2 3 + 5 7]

Out[90]= {{2 3, 6}, {5 7, 35}, 6 + 35, 41}
```

If you do not want to see all the intermediate steps, you can limit the intermediate steps recorded by specifying a second argument to `Trace`. The second argument can be either a symbol, or a pattern, or an option.

```
In[91]:=  Trace[2 3 + 5 7, Plus]

Out[91]=  {6 + 35, 41}

In[92]:=  Trace[2 3 + 5 7, Times]

Out[92]=  {{2 3, 6}, {5 7, 35}}

In[93]:=  Trace[2 3 + 5 7, x_ y_]

Out[93]=  {{2 3}, {5 7}}

In[94]:=  Trace[2 3 + 5 7, TraceDepth -> 1]

Out[94]=  {6 + 35, 41}
```

Input

When an `Input` statement is executed, you can type `?`*var* to obtain the value of the variable *var*.

```
In[95]:=  median[list_List] :=
              Module[{
                        sl,
                        len
                     },
                     len = Length[list];
                     sl = Sort[list];
                     Print[Input[ ]];
                     If[
                         OddQ[
                             Length[sl]
                         ],
                         sl[[(len + 1)/2 ]],
                         (sl[[len/2]] + sl[[len/2+1]])/2
                     ]
              ]
```

The variable `len` is local to the procedure. If you insert an `Input` statement in the middle of the procedure, *Mathematica* requests input from the user before calculating the median. At that time, you can ask *Mathematica* to provide the value of a variable such as `len`. In the following example, `len` has the value 3.

```
In[96]:=  median[{4, 5, 6}]
```

Verify Results

It is advisable to verify your results. *Mathematica* makes it easy to obtain results. Take time to check the results that you obtain. When possible, calculate results by alternative means.

Finding the inverse is often a good way to check results. For example, if you integrate an expression, try differentiating the result to see whether you get back the original expression. If you find the inverse of a matrix, check that the product of the matrix and its inverse returns the identity matrix, or something close to it (as rounding errors occasionally prevent exact numerical agreement).

General Advice

Finally, we repeat some insightful advice offered by Andrew Koenig in Chapter 8 of his book *C Traps and Pitfalls*, Addison–Wesley, 1989. The advice pertains just as well to any programming language.

Make your intentions plain.

Use parentheses liberally to make your intent clear. For example, `a+b*c` is not the same as `(a+b)*c`.

Look at trivial and extreme cases.

Make sure your code works on trivial cases and on unusual inputs. For many functions taking numeric arguments, special cases often include zero, one, and infinity.

Program defensively.

When you write programs that will be used by others, do not rely on platform-dependent aspects (e.g., color screens or large amounts of memory).

Keep an eye on the newsgroups.

Monitor the newsgroup `sci.math.symbolic` or join a *Mathematica* users group to keep abreast of other people's bug reports and advice.

15 Laboratories

The Ant and the Blade of Grass

Objectives: In this lab, you will learn how to use *Mathematica* to differentiate a function and construct a tangent line, to plot a series of points connected by line segments, to plot several things on the same graph, and to solve an equation.

Before Lab: Read Chapters 1, 2, and 3 of this manual.

***Mathematica* Commands:** You will need to differentiate functions using `D`, construct a tangent line, plot points along with functions, and solve equations using `Solve`. Before you start the lab, be sure you try each of the following examples.

- Differentiating functions (see Section 3.2).
  If a function f is defined by

```
In[1]:=  f[x_] := x^3
```

then its derivative is computed using *Mathematica*'s `D` operator, as follows.

```
In[2]:=  D[f[x], x]
```

Note that the output does not have a name. If you wish to give it a name, say Df, then you type

```
In[3]:=  Df = D[f[x], x]
```

- Tangent lines (see Section 3.2).
  If you want to find the line tangent to the curve $y = f(x)$ at the point $(a, f(a))$, you first find the slope $m = f'(a)$ and then use the point-slope equation of the line: $y - f(a) = m(x - a)$. By rearranging this equation and using the fact that $m = f'(a)$, we see that the equation of the tangent is given by $y(x) = f'(a) * (x - a) + f(a)$. You can make *Mathematica* find the right-hand side of this equation as follows:

```
In[4]:=  EquationOfTangent[f_, x_, a_] :=
           Module[{m  = D[f,x] /. x -> a,
                   fa = f      /. x -> a},
             Solve[y - fa == m (x - a), y][[1,1,2]]
           ]
```

You can check that this program is correct by trying it out on an arbitrary function:

```
In[5]:=  Clear[f, x, a]
         EquationOfTangent[f[x], x, a]

Out[6]=  f[a] - a f'[a] + x f'[a]
```

For the function $f(x) = x^3$ discussed previously, the slope of the tangent line at $x = 2$ is found from

```
In[7]:=  EquationOfTangent[x^3, x, 2]

Out[7]=  -4 (4 - 3 x)
```

Note that this is an expression without an assignment (there is no =), so nothing has been stored in memory. By assigning

```
In[8]:=  ftan[x_] := EquationOfTangent[x^3, x, 2]
```

you can also define the tangent line as a function (`ftan`). To see its formula, type

```
In[9]:=  ftan[x]
```

- Plotting several functions (see Sections 1.4 and 2.4).
  The function $f(x)$ and its tangent line $f_{\tan}(x)$, defined previously, can be plotted together on the interval $-5 \leq x \leq 5$ by enclosing the two functions in curly braces $\{\ \}$ and using the *Mathematica* `Plot` command.

```
In[10]:=  Clear[f, x, a];
          f[x_] = x^3;
          a = 2;
          Plot[
             Evaluate[{x^3,EquationOfTangent[x^3,x,2]}],
              {x,-5, 5}];
```

- Lists and plotting points (see Section 2.4).
  In *Mathematica* an ordered list is denoted by separating the items by commas and enclosing the list in curly brackets $\{\ \}$. Thus, the point $(2, 5)$ is entered as the ordered pair `{2,5}`, and a list of points might be

```
In[14]:=  octagon={{1,0},{4,0},{5,1},{5,4},{4,5},{1,5},
                   {0,4}, {0,1},{1,0}}
```

If you plot this list of points and connect the dots, you should get an octagon.

```
In[15]:=  ListPlot[octagon, PlotJoined -> True]
```

If you want to plot several lists of points simultaneously, you must first load in a special *Mathematica* graphics package called "Graphics`MultipleListPlot`". Do this and try the following:

```
In[16]:=  Needs["Graphics`MultipleListPlot`"];
          line1 = {{1,3},{2,3}};
          line2 = {{3,3},{4,3}};
          you = {{1,2},{2,1},{3,1},{4,2}};
          MultipleListPlot[octagon, line1, line2, you,
                                PlotJoined -> True]
```

What did you get?

Finally, you can even plot lists of points together with functions and also specify both the x-range and the y-range. For example,

```
In[21]:=  f[x_] := x^3;
          plot1 = Plot[Evaluate[{f[x],
                         EquationOfTangent[f[x], x, 2]}],
                         {x,0,5}];
          plot2 = MultipleListPlot[octagon, you,
                          PlotJoined->True];
          Show[plot1, plot2, PlotRange -> {0,20}]
```

- Solving equations (see Section 2.3).
  Suppose that you want to solve the equation $x/\pi + \sin(x) = 1$. You first enter the equation to make sure you have typed it properly.

```
In[25]:=  equation = x/Pi + Sin[x] == 1

Out[25]=  x
          -- + Sin[x] == 1
          Pi
```

You can then plot the left-hand and right-hand sides of this equation, to see where they intersect. To make things easier, first define a couple of functions that access the left- and right-hand sides of an equation.

```
In[26]:=  Lhs[x_ == y_] := x
          Rhs[x_ == y_] := y
          Plot[{Lhs[equation], Rhs[equation]},
              {x, -2Pi, 4Pi}]
```

So, there are three intersection points, i.e., solutions: one between 0 and 2, one between 2 and 4, and one between 4 and 6. To find the solutions, you first try *Mathematica*'s `Solve` command.

```
In[29]:=  Solve[eq, x]
```

`Solve` is unable to isolate x and hence solve this equation. So you have to resort to a numerical procedure. Unfortunately, this equation is not a polynomial so you cannot use `NSolve` either. Instead, you have to use *Mathematica*'s `FindRoot` command with different starting guesses to find approximate solutions.

```
In[30]:=  FindRoot[equation, {x, 0}]

Out[30]=  {x -> 0.827859}

In[31]:=  FindRoot[equation, {x, 3}]

Out[31]=  {x -> 3.14159}

In[32]:=  FindRoot[equation, {x, 5}]

Out[32]=  {x -> 5.45533}
```

We now have all three solutions.

Lab Report Requirements:

An ant is walking (to the right) over its ant mound, whose height (in inches) is given by the function:

$$h(x) = \frac{x^2/16 - 2x + 80}{(x^2/16 - 2x + 20)^2}$$

Nearby there is a blade of grass, which is located as the line segment from $(32, 1/5)$ to $(32, 8)$. The goal in this lab is to find the point where the ant first sees the blade of grass. You can assume that the ant's line of sight is the tangent line to the ant mound. The following series of questions will lead you to the solution. Answer them on the lab report form.

1. Define the function `h` that gives the height of the ant mound. Define the line segment occupied by the blade of grass and name it `grass`. Plot the ant mound and the blade of grass on the same graph.

2. Compute the derivative of `h` and name it `Dh`. On the lab report, record your *Mathematica* input but not the output.

3. Compute the tangent line to $y = h(x)$ at $x = 12.5$ and define it as a function `htan`. Plot the ant mound, the blade of grass, and the ant's line of sight when the ant is at $x = 12.5$. Can it see the blade of grass? Find the height H where the tangent line crosses the line $x = 32$ by evaluating `htan[32]`.

4. Compute the tangent line to $y = h(x)$ at $x = 15.5$ and define it as a function `htan`. Plot the ant mound, the blade of grass, and the ant's line of sight when the ant is at $x = 15.5$. Can it see the blade of grass? Find the height H where the tangent line crosses the line $x = 32$ by evaluating `htan[32]`.

5. We can now see that, when the ant is at some position $x = a$ between 12.5 and 15.5, it can first see the top of the blade of grass. We want to find a. So, compute the tangent line to $y = h(x)$ at $x = a$ for a variable a. Define the tangent line at $x = a$ as a function `htan`.

   Note: If you previously gave `a` a value, clear it by executing

   ```
   Clear[a]
   ```

6. You can no longer plot the tangent line because its formula contains a variable, namely, a. However, you can still find the height H where the tangent line crosses the line $x = 32$ by evaluating `htan[32]`. When this height H equals the height of the blade of grass, the ant can just begin to see the blade of grass. Use *Mathematica*'s `FindRoot` command to solve for the value of a where H equals the height of the blade of grass. You may need to specify a range for a in the `FindRoot` command. Denote the solution by A.

7. For the value A found in problem 6, compute the tangent line to $y = h(x)$ at $x = A$ and define it as a function `htan`. Plot the ant mound, the blade of grass, and the ant's line of sight when the ant is at $x = A$. Can it see the blade of grass? Find the height H where the tangent line crosses the line $x = 32$ by evaluating `htan[32]`.

8. *10% Extra Credit.* There is a second solution to the equation $H = 8$. What is wrong with this solution?

NAME_____________________ ID_____________________ Section_______

NAME_____________________ ID_____________________ Section_______

Lab: The Ant and the Blade of Grass

Lab Report

1. Complete the input lines:

   `h =`

   `grass =`

   Sketch your plot:

2. Complete the input line:

   `Dh =`

3. Equation of the tangent line at $x = 12.5$: ______________________________

   `htan =`

   Sketch your plot:

Can the ant see the grass?____________________

$H = h_{\text{tan}}(32) =$ ____________

4. Equation of the tangent line at $x = 15.5$: ____________________

```
htan =
```

Sketch your plot:

Can the ant see the grass? ____________

$H = h_{\text{tan}}(32) =$ ____________

5. Equation of the tangent line at $x = a$: ____________________

```
htan =
```

6. $H = h_{\text{tan}}(32) =$ ____________________________

   For your solve command: complete the input and write the output:

```
A = FindRoot[...]
```

7. Equation of the tangent line at $x = A$: ____________________

```
htan =
```

Sketch your plot:

Can the ant see the grass? ______________

$H = h_{\tan}(32) =$ ______________

8. *10% extra credit.* Answer in sentences and diagrams in the space below.

Calculus 1 Review

Objectives: In this lab, you will review the chain rule, differentiation, and integration, as well as the *Mathematica* commands for these operations.

Before Lab: Review: domain of a function, composition of functions, and the geometry of derivatives and integrals.

Throughout this lab, the symbols $f(x)$, $g(x)$, $h(x)$, and $k(x)$ will refer to the following functions

$$f(x) = \sqrt{x-2}\,, \quad g(x) = x^2\,, \quad h(x) = f \circ g(x)\,, \quad k(x) = g \circ f(x)$$

You will be doing various computations with these four functions, so you should enter them now. The *Mathematica* commands for entering the functions $f(x)$ and $g(x)$ are

```
In[1]:=  f[x_] := Sqrt[x - 2]
         g[x_] := x^2
```

Note: There are three ways of entering the composition of two functions in *Mathematica*. Thus, one could enter the function $k(x)$ with any of the following *Mathematica* commands.

```
In[3]:=  f @ g[x]

Out[3]=  f[g[x]]

In[4]:=  Composition[f,g][x]

Out[4]=  f[g[x]]

In[5]:=  f[g[x]]

Out[5]=  f[g[x]]
```

To see the explicit algebraic expression of $k(x)$, for example, enter the *Mathematica* command

```
In[6]:=  k[x]
```

Lab Report Requirements: Answer the following questions on the lab report form.

1. Using the *Mathematica* command `Plot[{f(x),g(x)},{x,-7,7}]`, plot the graphs of $f(x)$ and $g(x)$ for x between -7 and 7. Reproduce this graph on the lab report form, and be sure to label which graph belongs to which function.

2. The function g has all real numbers for its domain. What is the domain for the function f?

3. Explain why the plot for $f(x)$ is not drawn for all numbers between -7 and 7.

4. Plot the graphs of $h(x)$ and $k(x)$ in *Mathematica* as in problem 1, and then copy these plots to the lab report form. Be sure to label the graphs.

5. Explain why the straight line passing through the point $(0, -2)$ is not the graph of $k(x)$, and draw a better graph.

6. Without using *Mathematica*, compute the derivatives of the two functions $h(x)$ and $k(x)$. After you have done this, compare your answer with *Mathematica*'s.

   Note: You can have *Mathematica* calculate derivatives using the `D` operator.

```
derivk = D[k[x], x]
```

7. The answer you got for the derivative of $k(x)$ should *not* be the same as *Mathematica*'s. Why not?

8. Sketch the region bounded by the curves $y = 0, y = f(x), y = g(x)$, and $x = 6$.

9. Calculate the area bounded by the curves you sketched in problem 8, by using the *Mathematica* commands

```
Integrate[g[x], {x,0,2}] - Integrate[g[x]-f[x],{x,2,6}]
```

   Explain what each of the integrals represents.

   Note: Mathematica does not echo back what you type in as input; it only prints out the answers. Get in the habit of checking the arguments in complicated inputs by first assigning, them to symbols and then using these symbols to stand for the expressions. When *Mathematica* makes such an assignment, it will output a formatted version of your expression. This makes it a lot easier to find and correct typos or other mistakes.

10. Explain the differences in output between the following approaches to calculating the derivative of the function $\frac{1}{x(x-2)}$.

    (a) `D[1/x (x-2), x]`

    (b) `p = (x (x-2))^(-1); D[p, x]`

(c) `D[1/(x (x-2)), x]`

(d) Which values are correct? How can the others be fixed? After fixing them, which approach do you feel is best? Why?

NAME_____________________ ID_____________________ Section_______

NAME_____________________ ID_____________________ Section_______

Lab: Calculus 1 Review

1.

2. Domain = ____________________

3.

4.

5.

6. $h'(x) =$_____________________ $k'(x) =$_____________________

7.

8.

9. Area = ______________________

Explanation:

10.

11. Discussion:

The Chain Rule

Objectives: The purpose of this drill is to help you learn to compute derivatives using the chain rule.

Prerequisites: It is assumed that you have seen the chain rule in class or read about it in your book.

***Mathematica* Commands:** You are not expected to know any *Mathematica* commands except how to enter an assignment and how to execute a command.

Initialization:

1. Open the *Mathematica* Notebook called `chain.ma` by double-clicking on its icon. The Notebook `chain.ma` may be found in the directory `chap15/chain`. Your instructor will tell you how to find this file.
2. All the functions needed to run the drill are automatically loaded as the file opens. No external functions need be loaded.

Procedure:

1. Execute the command `StartChainRuleDrill[]` by putting the cursor on this line and pressing ⟨SHIFT⟩ and ⟨RETURN⟩ simultaneously. A window will pop up that contains a question asking you to differentiate a particular composite function $F(x) = f(g(x))$.
2. Below the question is a box containing a blinking cursor. This is the answer window. Work out your answer on paper and then type it in the answer window. *Mathematica* will then check your answer and tell you whether you are right or wrong. In the unlikely event that you are wrong, *Mathematica* will ask you if you want to try again. You can try as many times as you like.
3. If you need help, just type "help" in the answer window. *Mathematica* will print out a step-by-step explanation of how to tackle the problem.
4. If you are utterly stuck, you can ask *Mathematica* to give you the answer by typing "show me" in the answer window. You can stop the drill at any time by typing "quit."
5. Remember that the derivative of a composite function $f(g(x))$ is $f'(g(x))g'(x)$.

Lab Report Requirements: If this drill is done as a lab, run through the drill as many times as necessary until you feel confident in your ability. You will be asked to prove proficiency by either turning in a lab report or by taking a quiz.

As a lab report, go through the drill for three different composite functions and fill out the attached answer sheet.

As a quiz, you will be given a copy of the attached answer sheet with three composite functions filled in. You answer the rest of the questions.

NAME________________________ ID________________________ Section________

NAME________________________ ID________________________ Section________

Lab: The Chain Rule

Lab Report

Run 1	Run 2	Run 3
Practice function $F(x) = f(g(x))$		
Inner function $u = g(x)$		
Outer function $f(u)$		
Deriv. of outer function $Df(u) = df/du$		
Deriv. of outer function with u replaced by $g(x)$		
Deriv. of inner function $Dg(x) = dg/dx$		
Deriv. of comp. function $DF(x) = dF/dx$		

Compound Interest

Objectives: You will get practice computing simple and compound interest and learn about continuous compounding. In the process, you will discover the limit of an important sequence of numbers.

Before Lab: Read the entire lab and do problems 1, 2(a-c), 3, 4, and 5a. Feel free to use a calculator. You will redo them in lab using *Mathematica*.

***Mathematica* Commands:** You will need to use the *Mathematica* commands that define and use functions of several variables and compute limits at infinity. Examples are given here.

- Functions of several variables(see Section 2.2).
  The following *Mathematica* command defines a function `distance`, which gives the distance between the point (a, b) and the point (c, d).

  ```
  distance[a_,b_,c_,d_]:= Sqrt[(a-c)^2 + (b-d)^2]
  ```

  Then, the distance between $(2, 3)$ and $(5, 7)$ can be found from

  ```
  distance[2,3,5,7]
  ```

  Try it!

- Limits at infinity see Section 1.5.
  If the function $f(x) = \dfrac{2x^2 - 3x}{1 - x^2}$ is defined by the *Mathematica* command

  ```
  f[x_] := (2 x^2 - 3 x)/(1 - x^2)
  ```

  then its limit as x goes to infinity is computed using

  ```
  In[1]:=  Limit[f[x], x -> Infinity]
  Out[1]=  -2
  ```

  Try it!

Lab Report Requirements: Answer the following questions on the lab report form.

1. Suppose you put \$1000 in a bank that pays 6% simple interest paid out annually.

   ***Mathematica* Procedure:** *Use Mathematica to do your arithmetic, as you would a calculator. There are no formulas or functions to define.*

   (a) How much interest will you receive at the end of 1 year?

   (b) If you redeposit your \$1000 plus the interest, how much money will you have in the bank at the end of 1 year?

   (c) If you leave your money in the bank for a second year, the money you have at the beginning of the second year will receive 6% simple interest. How much interest will you receive at the end of the second year?

   (d) If you redeposit your principal and interest, how much money will you have in the bank at the end of the second year?

   When you receive 6% simple interest for each year and redeposit the principal and interest at the end of each year, we say that you are receiving 6% interest compounded yearly.

2. Suppose you put $\$P$ (principal) in a bank that pays interest at a rate r compounded yearly. (If the rate is 5%, then $r = .05$.)

   ***Mathematica* Procedure:** *Start by repeating your computations from problem 1, but replace 1000 by P and .06 by r. Do not use any formulas or functions until part (e).*

   (a) In terms of P and r, how much money will you have in the bank at the end of 1 year?

   (b) After 2 years?

   (c) After t years? Why?

   Here is the first answer to show what is expected. Copy it to your answer sheet. Hereafter, you are expected to answer on you own.

   Answer:

   ```
   A = P (1 + r)
   ```

   For each 1 year period, the amount in the bank is multiplied by a factor of $(1 + r)$. Multiplying the previous amount by 1 indicates this money is still in the bank. Multiplying by r gives the interest for the year. Adding these together gives the principal at the end of the year.

(d) Express your answer as a *Mathematica* function A of the three variables P, r, and t.

Answer:

```
A[P_,r_,t_] := P (1 + r)^t
```

(e) Use your *Mathematica* function A to compute the amount of money in the bank if \$2000 is invested at 4% interest compounded annually for 6 years.

```
A[2000, .04, 6]
```

(f) Use your *Mathematica* function A to compute the amount of money in the bank after 4 years if \$2000 is invested at 6% interest compounded annually.

3. Suppose you put \$1000 in a bank that pays 6% interest compounded semi-annually (i.e., twice a year).

   ***Mathematica* Procedure:** *Once again, use Mathematica to do your arithmetic. Do not use any formulas or functions.*

   (a) How much is in the bank after 6 months? *(Ask your instructor to check your answer!)*

   (b) After 1 year?

   (c) After 2 years?

4. Suppose you put \$$P$ in a bank that has an interest rate r compounded semi-annually.

   ***Mathematica* Procedure:** *Start by repeating your computations from problem 3, but replace 1000 by P and .06 by r. Do not use any formulas or functions.*

   (a) In terms of P and r, how much money will you have in the bank at the end of 1 year?

   (b) After 2 years?

   (c) After t years? Why? *Hint:* Factor you answer to (b). Be sure to try your formula for $t = 2$.

5. Suppose you put \$$P$ in a bank that has an interest rate r compounded n times a year.

(a) In terms of P, r, t, and n, how much money will you have in the bank at the end of t years? Why?

***Mathematica* Procedure:** *There is nothing to compute. Simply generalize your formula from problem 4c.*

(b) Express your answer as a *Mathematica* function A of the four variables P, r, t, and n.

(c) Use your *Mathematica* function A to compute the amount of money in the bank if \$3000 is invested for 10 years at 5% interest compounded monthly.

6. Suppose you put \$1 in a bank that pays 100% interest ($r = 1.00$) compounded n times a year. *(Obviously, this interest rate is unrealistic, but it will illustrate a point.)*

***Mathematica* Procedure:** *Use your formula from problem 5b. After typing part (a), for each later part simply copy, paste, and edit your input from part (a). This will save time.*

How much money will you have in the bank after 1 year, if it is compounded

(a) semi-annually?

(b) quarterly?

(c) monthly?

(d) weekly? (Assume 52 weeks in a year.)

(e) daily? (Assume 365 days in a year.)

***Mathematica* Procedure:** *Wrapping* `N` *with a second (integer) argument around an expression* `expr` *such as*

```
In[2]:=  N[expr, 20]
```

causes the expression to be calculated to 20 digits' precision. You may need this tip to continue the lab.

(f) hourly?

***Mathematica* Procedure:** *If the computation is taking too long, try making the rate a decimal:* $r = 1.0$.

(g) every minute?

(h) every second?

(i) continuously?

***Mathematica* Procedure:** *Use Mathematica's* `Limit` *command.*

7. Write the answer to the previous problem in terms of known mathematical constants.

   You will be able to derive this limit once you learn l'Hôpital's rule.

8. *Now let's be realistic:* Suppose you put $1000 in a bank that pays 3% interest compounded continuously.

   (a) How much would you have at the end of 1 year?

   ***Mathematica* Procedure:** *Use your formula from problem 5b and Mathematica's* `Limit` *command.*

   (b) Compute the number $e^{.03}$. *Note:* In *Mathematica*, the exact value of the constant e is entered by typing `E`.

   (c) Compare your answers for parts (a) and (b).

9. Suppose you put $\$P$ in a bank that has an interest rate r compounded continuously.

   (a) How much do you think you will you have in the bank after 1 year? Why?

   (b) What answer does *Mathematica*'s `Limit` command give you?

   (c) How much do you think you will you have in the bank after t years? Why?
   Give your answer in terms of e.

   (d) What answer does *Mathematica*'s `Limit` command give you?

NAME______________________ ID______________________ Section_______

NAME______________________ ID______________________ Section_______

Lab: Compound Interest

Lab Report

1. (a) Interest = _______________

   (b) Amount = _______________

   (c) Interest = _______________

   (d) Amount = _______________

2. (a) A = _______________

   (b) A = _______________

   (c) A = _______________

   Reason:

   (d) A = _______________

   (e) A = _______________

   (f) A = _______________

3. (a) A = _______________

   (b) A = _______________

   (c) A = _______________

4. (a) A = _______________

   (b) A = _______________

   (c) A = _______________

   Reason:

5. (a) A = _______________

   Reason:

(b) A := ______________

(c) A = ______________

6. (a) A = ______________ (semi-annually)

   (b) A = ______________ (quarterly)

   (c) A = ______________ (monthly)

   (d) A = ______________ (weekly)

   (e) A = ______________ (daily)

   (f) A = ______________ (hourly)

   (g) A = ______________ (every minute)

   (h) A = ______________ (every second)

   (i) A = ______________ (continuously)

7. A = ______________

8. (a) A = ______________

   (b) $e^{.03} =$ ______________

   (c) Discussion:

9. (a) A = ______________

   Reason:

   (b) A = ______________

   (c) A = ______________

   Reason:

   (d) A = ______________

Graphic Drill on Derivatives and Second Derivatives

Objectives: The purpose of this drill is to help you develop an intuitive understanding of the relationship between a function, its derivative, and its second derivative. You will be asked two kinds of questions. Sometimes, you will be shown the graph of a function f together with those of its first and second derivatives, f' and f'', respectively. Each curve will be shown in a separate color. You will then be asked to identify which color goes with which curve purely on the basis of the shapes of the curves. At other times, you will be shown the graph of a function and told its definition and then asked questions about the coordinates of local maxima, local minima, and inflection points.

Prerequisites: It is assumed that you already know the definitions of the following terms: increasing, decreasing, local maximum, local minimum, concave up, concave down, and inflection point.

***Mathematica* Commands:** You are not expected to know any *Mathematica* commands except how to enter an expression and how to execute a command.

Initialization:

1. Open the *Mathematica* Notebook called `d_dd.ma` by double clicking on its icon. It is in the `d_dd` subdirectory of the `chap15` directory. Your instructor will tell you how to find this file.

2. As the file opens, all the necessary definitions needed to run the drill are automatically loaded into *Mathematica*.

3. We should warn you that as you run the drill, a new window will appear in which you will type your answers. You should reposition this window by dragging it so that you can view both the Notebook and the answer window simultaneously.

Procedure:

1. To run the drill, execute the command `StartDrill[]` by putting the cursor on this line and pressing ⟨SHIFT⟩ and ⟨RETURN⟩ simultaneously. A graph will appear with some explanatory text below it and an answer window will pop up. This window contains the question you are to answer and a box containing a blinking cursor in which your are to type your answer.

2. Sometimes you will be shown a plot consisting of a red curve, a green curve, and a blue curve corresponding (in some order) to the graphs of a function

$f(x)$ and its first and second derivatives, f' and f'' respectively. Your job is to guess which color corresponds to which function. The color permutation is randomly selected in each question. The drill will prompt you for a color for each function in sequence. In each case, you should answer "red," "green," or "blue."

3. If the computer shows you the graph of a single function, it will also tell you its equation and prompt you for coordinates corresponding to the location of local maxima, local minima, or inflection points. In this case your answers will be coordinates expressed in *Mathematica* syntax (that is, as a list of pairs delimited by curly braces. So the point (0,1) should be entered as $\{0, 1\}$).
4. *Mathematica* will check your answers and give you feedback.
5. If you want to use *Mathematica* to help you work out the answers, you can either open a separate *Mathematica* Notebook and work in that or answer `InterruptDrill[]` to any question. This will allow you to run *Mathematica* commands in your Notebook. When you have finished and feel ready to answer the question, you must then type `Return[]`. This will take you back to where you left off.
6. If you have trouble, you can enter "hint" to get a hint or "show me" to see the answer.
7. If you are asked, say, for the coordinates of inflection points, and there are none, reply with a period, "."
8. When you have finished with the questions about a given graph, delete all plots and then reexecute `StartDrill[]` to get a new original graph.

How to Enter an Answer

The following are examples of possible answers.

```
Answer = -7.34
Answer = {-7.34, 10.00}
Answer = {-7.34, 2.57, 8.31}
Answer = {{-7.34, 2.57}, {8.31, 10.00}}
Answer = Null
Answer = 2
```

1. The answers will be in the form of one or more x-coordinates or intervals or a plot number; in some cases, there might be no answer.

2. The x-coordinates in your answers must be chosen from the list given to you with the graph. All x-coordinates must be entered with two decimal places. Thus, 2.30 and 4.00 are acceptable, but 2.3, 4, 4., and 4.0 are unacceptable.

3. If the answer or a part of an answer is an interval, enter it as $[a, b]$, where a and b are x-coordinates taken from the list. This notation is not meant to indicate whether the endpoints are or are not included.

4. If there is more than one number or interval for an answer, separate them by commas.

5. If there is no answer to the question, then type `Null` For example, if the question asks for the location of all local minima, and there are no local minima, then `Null` is the correct answer.

6. If the question is to identify a plot, just type the number of the plot (and a semicolon).

Lab Report Requirements: If this drill is done as a lab, run through the drill as many times as necessary until you feel confident in your ability. You will be asked to prove proficiency by either turning in a lab report or taking a quiz.

As a lab report, go through the drill for three different functions: once where you are shown the graph of f, once where you are shown the graph of f', and once where you are shown the graph of f''. As you answer the questions, fill out the accompanying answer sheet once for each function. You will be asked to give the following information.

1. For the original graph shown to you, include
   (a) a rough sketch of this plot;
   (b) a statement of whether this plot is f or f' or f'';
   (c) a list of the numbers to be used in the answers.

2. Then for each question, include
   (a) the statement of the question (including a rough sketch of any plots shown);
   (b) your answer;
   (c) a sentence saying how you arrived at your answer.

As a quiz, you will be given a copy of the answer sheet with the questions and plots filled in. Answer the questions using complete sentences and describe how you arrived at your answers.

NAME____________________ ID____________________ Section______

NAME____________________ ID____________________ Section______

Lab: Graphic Drill on Derivatives

Lab Report

CIRCLE: Graph of f or f' or f'':
Numbers to be used in the answers:

Question 1:
Answer 1:
Explanation:

Question 2:
Answer 2:
Explanation:

Question 3:
Answer 3:
Explanation:

Question 4:
Answer 4:
Explanation:

Question 5 (if any):
Answer 5:
Explanation:

Plots shown for last two questions:

Visualizing Euler's Method

Objectives: The goal of this lab is to use *Mathematica* to illustrate Euler's method, and to show how Euler's method leads naturally to numerical antidifferentiation, numerical integration, and linear splines.

Background: Euler's method for approximating the solution to the problem

$$\frac{dy}{dt} = f(t, y), \quad y(t_0) = y_0 \tag{1}$$

is as follows: choose a step size $h > 0$, and let $t_i = t_0 + ih$, $i = 0, 1, 2, ..., n$, where n is the number of steps of size h you must take to reach some prescribed final value of t. Then, recursively compute the values

$$y_{i+1} = y_i + h\, f(t_i, y_i), \quad i = 0, 1, 2, ..., n-1. \tag{2}$$

The y_i's are intended to be approximations to the true solution, $y(t)$, at the discrete points t_i; that is, $y_i \approx y(t_i)$ for $i = 0, 1, ..., n$. To see where the method comes from, observe that the derivative of the true solution at t_0 is $y'(t_0) = f(t_0, y(t_0)) = f(t_0, y_0)$, as follows from Equation (1). Hence, the line tangent to the graph of $y(t)$ at the point (t_0, y_0) has the equation $y = y_0 + (t - t_0)f(t_0, y_0)$. If we use this as an approximation to the true solution for t near t_0, then when we let $t = t_0 + h = t_1$, we obtain the approximation $y_1 = y_0 + h\, f(t_0, y_0)$ for $y(t_1)$. This is Equation (2) with $i = 0$. The algorithm in Equation (2) is just a continuation of this procedure.

We show one way of implementing Euler's method with *Mathematica*, using a procedure called `Euler`. The procedure is as follows. The output is a list containing the values $y_0, y_1, ..., y_n$.

```
Euler[{D[y_[t_],t_]==f_, y_[t0_]==y0_}, {t_,t0_,t1_}, n_]:=
    Module[{i,y$,h},
        h       = N[(t1-t0)/n];
        y$[0]   = y0;
        y$[i_]:= y$[i] = (y$[i-1] + h f //.
            {y[t]->y$[i-1],t->(t0+(i-1) h)});
        solution = Table[{i h, N[y$[i]]}, {i,0,n}];
        Clear[y$];
        solution
    ]
```

Note that to obtain the value of the solution at each successive step, the recursion relation requires the value at the previous step. Consequently, to make a more efficient program, we create a recursive function that "remembers" previous values. We do this by inserting into the usual *f* [*x*_] := *body* form of a function the clause *f*[*x*]= to get something like *f* [*x*_] := (*f*[*x*]= *body*). Any time the function *f* is called for a particular value of x, such as x = x0, the body is executed and the result assigned to f`[x0]`. Subsequent calls to f`[x0]` are obtained immediately by direct table lookup without any need for further computation.

The other trick we have introduced into our definition of Euler's method is the definition of the recursive function within the scope of some other function (the Euler function). This means that our recursive function is defined afresh each time the Euler function is called. Any remembered values are cleared out before the Euler function terminates. This is good programming practice for memory functions.

Our Euler function takes 3 arguments:

1. the first-order differential equation to be solved, together with an initial condition
2. the range over which the solution is sought
3. the number of steps to use.

The following example will illustrate how the `Euler` procedure is used in a typical problem, and how to plot the results. In Equation (1), we let $f(t, y) = y$, $t_0 = 0$, and $y_0 = 1$. The exact solution in this case is easily seen to be $y(t) = e^t$. We will also plot the direction field and the true solution, and display all three plots together. As this is only an example, we'll use just $n = 4$ steps. If you wanted a better numerical solution, you would use a much larger value of n.

```
numericalSoln =
    Euler[{D[y[t],t]==y[t],y[0]==1}, {t,0,1}, 4];
Show[ListPlot[numericalSoln,
        PlotJoined -> True,
        DisplayFunction -> Identity],
    ListPlot[numericalSoln,
        PlotStyle -> PointSize[0.02],
        DisplayFunction -> Identity],
    Plot[Exp[t], {t,0,1},
        DisplayFunction -> Identity],
    PlotVectorField[{Exp[t],1}, {t,0,1}, {y,0,3},
        DisplayFunction -> Identity],
    DisplayFunction -> $DisplayFunction]
```

We can take a look at this solution by entering:

```
ListPlot[numericalSoln, PlotJoined -> True]
```

We next include the direction field. To do this, we must first load the package `Graphics`PlotField`` using `Needs`.

```
Needs["Graphics`PlotField`"];
p1 = ListPlot[numericalSoln, PlotJoined->True,
        DisplayFunction -> Identity];
p2 = ListPlot[numericalSoln,
        PlotStyle -> PointSize[0.02],
        DisplayFunction -> Identity];
p3 = PlotVectorField[{Exp[t], 1}, {t,0,1}, {y,0,3},
        DisplayFunction -> Identity];
Show[p1, p2, p3,
        DisplayFunction > $DisplayFunction]
```

The purpose of this plot is to show how the straight-line segments in the Euler solution have slopes that agree with the direction field at the *left* endpoint of the line segment. You will be asked in the exercises to explain this. Finally, we include the true solution, to get an idea of how good the approximation is.

```
p4 = Plot[Exp[t], {t,0,1},
    DisplayFunction -> Identity];
Show[p1,p2,p3,p4,
    DisplayFunction -> $DisplayFunction
```

Exercises

1. Apply Euler's method with $n = 2$ to the problem

$$\frac{dy}{dt} = f(t,y) = 2\sin\left(1 + 4\,e^t\right) \equiv g(t), \quad y(0) = 0, \quad 0 \le t < 1 \qquad (3)$$

2. What is the exact solution to Equation (3)? *Hint:* It is okay to give your answer as an integral, with the variable upper limit t. How is this solution related to the set of antiderivatives of $g(t)$?

3. Use your Euler solution to obtain an approximation to the definite integral $\int_0^1 g(t)\,dt$. Compare this with the answer obtained using *Mathematica*'s `Integrate` command.

4. Plot the direction field and the Euler solution together, in the plot window $0 \le t \le 1, \quad -1 \le y \le 1$.

5. Find the piecewise linear function $s(t)$ that goes through the points $\{(t_i,\ y_i)\}_{i=0}^{n}$. For example, the piecewise linear function that goes through the points (0, 1), (.5, 1), (1, 0) is $s(t) = \begin{cases} 2t & \text{if } 0 \le t \le .5 \\ 2 - 2t & \text{if } .5 < t \le 1 \end{cases}$

   Next, use on-line help to find out how to use *Mathematica*'s `Interpolation` command to obtain $s(t)$, and compare your answer with *Mathematica*'s.

6. How are the slopes of the straight-line segments in the graph of the Euler solution related to the direction field? Explain the reason for this relation.

7. Repeat exercises 1, 3, and 4 with $n = 4$.

8. Euler's method is usually too inaccurate to be of practical use, but it is a good starting point for learning about numerical methods for solving differential equations. A significant improvement is the following, called the modified Euler's method (or Huen's method, or second-order Runge-Kutta method):

$$y_{i+1} = y_i + \frac{h}{2}(k_1 + k_2), \quad \text{where } k_1 = f(t_i, y_i) \text{ and } k_2 = f(t_i + h, y_i + hk_1)$$

   Write a procedure called `Eulermod` to implement this method, and apply it to the example (i.e., with solution $y = e^t$), using $n = 2$. Display plots showing the approximate solution, together with the exact solution. Repeat for $n = 4$.

Numerical Integration

Objectives: In this lab, you will investigate and compare five methods of computing integrals numerically.

Before Lab: Read Chapter 7. Note that several commands for numerical integration are not discussed in Chapter 7, but they are explained here. Also, read the section in your calculus textbook on numerical integration.

***Mathematica* Commands:** You will need to use *Mathematica*'s `NIntegrate` command to compute a numerical value of an integral. In addition, you will need to use several *Mathematica* commands that we have defined in the `numericalIntegration.ma` Notebook, namely: `LeftBox`, `RightBox`, `MiddleBox`,`TrapezoidBox`, `LeftSum`, `RightSum`, `MiddleSum`, `TrapezoidSum`, and `SimpsonSum`. Further, you should feel free to try animating these commands as the number of approximating boxes is increased. An example of how to do this is shown in the Notebook. Each of these commands will now be explained.

- `LeftSum` and `LeftBox`,

  Mathematica's built in `Sum` command can compute only the closed forms of very simple sums. However, `Sum` can be redefined and enhanced by loading the package ``Algebra`SymbolicSum``. After this package has been loaded, you will be able to obtain closed forms of a wide variety of sums.

  Assume a function $f(x)$ has been defined. The *Mathematica* command `LeftSum` defined by:

```
LeftSum[fn_, {x_, a_, b_}, n_]:=
    Module[{dx = (b - a)/n},
        Expand[
            Sum[(fn /. x -> (a + i dx)) dx,
                {i, 0, n-1}]
        ]
    ]
```

  will compute the Riemann sum approximation to $\int_a^b f(x)\,dx$ using n rectangles of equal width, where the height of each rectangle is the value of $f(x)$ at the left endpoint of the interval. The command `LeftBox` defined by

```
LeftBox[fn_, {x_, a_, b_}, nRectangles_]:=
    Show[
        Plot[fn, {x,a,b}, DisplayFunction->Identity],
        LeftSidesAndTops[fn,{x,a,b},nRectangles],
        DisplayFunction->$DisplayFunction
    ];

LeftSidesAndTops[fn_, {x_, a_, b_}, n_]:=
    Module[{dx, leftSides},
        dx = N[(b-a)/n];
        leftSides = Table[LeftSideAndTop[fn,a,dx,i],
                    {i,0,n-1}];
            Join[leftSides,
            {Graphics[{Line[{{b,0},{b, fn /.
                                  x->(b-dx)}}]}]}]
        ]
    ];

LeftSideAndTop[fn_, a_, dx_, i_]:=
    Module[{yBtmLeft, yTopLeft, yTopRight, yBtmRight},
        yBtmLeft  = {a + i dx, 0};
        yTopLeft  = {a + i dx, fn /. x->(a + i dx)};
        yTopRight = {a + (i+1) dx, fn /.  x->(a+i dx)};
        yBtmRight = {a + (i+1) dx, 0};
        Graphics[{Line[{yBtmLeft, yTopLeft}],
                  Line[{yTopLeft, yTopRight}],
                  Line[{yBtmRight, yTopRight}]
                  }
        ]
    ];
```

and invoked with:

```
LeftBox[f[x],{x,a,b},n]
```

will plot $f(x)$ along with these rectangles. Further, if you are working within a *Mathematica* Notebook, you can animate this plot with the command:

```
Map[LeftBox[f[x],{x,a,b},#]&, Table[2^i, {i, 2, k}]]
```

This command produces a sequence of frames of `LeftBox` plots with $n = 1, 2, 4, ..., 2^k$ rectangles.

- `RightSum` , `RightBox`, `MiddleSum`, and `MiddleBox`
  These commands are the same as `LeftSum` and `LeftBox`, except that the height of each rectangle is the value of $f(x)$ at the right endpoint or the midpoint of the interval.

- `TrapezoidSum` and `TrapezoidBox`,
  The command

  ```
  TrapezoidSum[f[x], {x, a, b}, n]
  ```

  will compute the trapezoid rule approximation to $\int_a^b f(x)\,dx$ using n trapezoids of equal width, where the heights of each trapezoid are the values of $f(x)$ at the two endpoints of the interval. The command

  ```
  TrapezoidBox[f[x], {x, a, b}, n]
  ```

  will plot $f(x)$ along with these trapezoids. Further, the command

  ```
  Map[TrapezoidBox[f[x], {x,a,b}, #]&, Table[2^i, {i,2,k}]]
  ```

  will produce a sequence of frames showing `TrapezoidBox` using $n = 1, 2, 4, ..., 2^k$ trapezoids. Double-clicking on any one of these frames will animate the sequence.

- `SimpsonSum`
  The command

  ```
  SimpsonSum[f(x),{x,a,b},n]
  ```

  will compute the Simpson's rule approximation to $\int_a^b f(x)\,dx$ using $\frac{n}{2}$ boxes of equal width, whose upper edge is a piece of the parabolathat matches the values of $f(x)$ at the two endpoints and the midpoint of the interval (n must be even).

Initialization: Find the `numericalIntegration.ma` Notebook on your computer. It is in the directory `chap15/numint`. Double-click on its icon. The Notebook will open and all the functions needed to do the exercises will automatically be loaded.

Lab Report Requirements: Answer the following questions by filling out the lab report form. Note that your results from problems 2c, 3c, 4d, 5d, and 6b should be entered in the tables at the end of the report.

1. We would like to study an integral that cannot be evaluated using standard integration techniques in terms of standard functions.

   (a) Define $f(x)$ to be one of the following four functions, using arrow notation. Your instructor will tell you which one to use.

$$f(x) = \frac{e^{x/2}}{x^3}, \quad f(x) = 2 - \frac{e^{x/2}}{x^3},$$
$$f(x) = 2\sqrt{x}\sin\left(\frac{x^2}{64}\right), \quad f(x) = 2 - 2\sqrt{x}\sin\left(\frac{x^2}{64}\right)$$

   (b) Plot $f(x)$ with the ranges $\{x, 1, 5\}$ and $\{y, 0, 2\}$ and determine whether it is increasing or decreasing and concave up or concave down on this interval.

   (c) Use *Mathematica*'s `Integrate` and `NIntegrate` commands to compute $\int_1^5 f(x)\,dx$. Note that *Mathematica* cannot compute the the answer in terms of elementary functions. Save the decimal answer as a *Mathematica* variable `A`.

2. We would like to understand how *Mathematica* might find a numerical approximation to this integral.

   (a) Use the `LeftBox` command to plot $f(x)$ on the interval $[1, 5]$ together with eight rectangles for the left Riemann sum approximation. Put a rough sketch in your lab report.

   (b) Will this approximation produce an upper bound or a lower bound for the integral? Why? *Hint:* Does this depend on the function being increasing or decreasing and/or concave up or concave down?

   (c) Use the `LeftSum` command to compute the left Riemann sum approximation using $n = 4, 16, 64$, and 256 rectangles.

   ***Mathematica* Procedure:** *Copy, paste, and edit it for the other three times.*

   Also compute the error with respect to the actual integral, A. (The error should be positive if the approximation is too large and negative if too small.) Enter your results in the tables at the end of the lab report, keeping at least three decimal places.

3. Repeat problem 2 using `RightBox` and `RightSum` to consider the right Riemann sum approximation.

   Mathematica **Procedure:** *Copy, paste, and edit your commands from problem 2.*

4. Your answers to problems 2b and 3b should show that either the left or right Riemann sum produces a lower bound and the other produces an upper bound. So, the average of these two methods must give a better estimate of the integral.

   (a) Explain why the trapezoid rule gives the average of the left and right Riemann sum approximations.

   (b) Use the `TrapezoidBox` command to plot $f(x)$ on the interval $[1, 5]$ together with four trapezoids for the trapezoid rule approximation.

   Mathematica **Procedure:** *Copy, paste, and edit your commands from problem 2.*

   (c) Will this approximation produce an upper bound or a lower bound for the integral? Why? *Hint:* Does this depend on the function being increasing or decreasing and/or concave up or concave down?

   (d) Use the `TrapezoidSum` command to compute the trapezoid approximation using $n = 4, 16, 64,$ and 256 trapezoids. Also compute the error with respect to the actual integral, A. Enter your results in the tables.

5. Another result that is better than the left and right Riemann sum approximations may be obtained by taking the height of the rectangles to be the value of $f(x)$ at the midpoint of the interval.

   (a) Explain why the midpoint Riemann sum gives a better approximation than the left and right Riemann sums. *Hint:* Does this depend on the function being increasing or decreasing and/or concave up or concave down?

   (b) Use the `MiddleBox` command to plot $f(x)$ on the interval $[1, 5]$ together with four rectangles for the midpoint Riemann sum approximation.

   (c) Will this approximation produce an upper bound or a lower bound for the integral? Why? *Hint:* Does this depend on the function being increasing or decreasing and/or concave up or concave down?

   (d) Use the `MiddleSum` command to compute the midpoint Riemann sum approximation using $n = 4, 16, 64,$ and 256 rectangles. Also compute the error with respect to the actual integral, A. Enter your results in the tables.

6. Your answers to problems 4c and 5c should show that either the trapezoid rule or the midpoint rule produces a lower bound and the other produces an upper bound. So, an average must give a better estimate. However, you should notice that the error from the midpoint rule is about half of the error from the trapezoid rule. Simpson's rule effectively consists of two-thirds of the midpoint rule and one-third of the trapezoid rule, thus giving the midpoint rule twice as much weight.

   (a) *20% extra credit.* Explain why Simpson's rule is related to the trapezoid and midpoint rules by the formula: $S_{2n} = 1/3T_n + 2/3M_n$ (where the subscript indicates the number of intervals).

   (b) Use the `SimpsonSum` command to compute the Simpson's rule approximation for $n = 4, 16,\ 64,$ and 256 intervals. Also compute the error with respect to the actual integral, A. Enter your results in the tables.

7. *20% extra credit.* Error Analysis. The error in each of the five methods of approximating the integral is known to be approximately proportional to $(\Delta x)^k$, with a different k for each method. Here

$$\Delta x = \frac{b-a}{n} = \frac{5-1}{n} = \frac{4}{n}$$

   Use the table to find the k for each method. Explain the results.

NAME____________________ ID____________________ Section_______

NAME____________________ ID____________________ Section_______

Lab: Numerical Integration

Lab Report

Enter your results from problems 2(c), 3(c), 4(d), 5(d), and 6(b) in the tables at the end.

1. The integral:

   (a) $f(x) =$ ______________________________

   (b) Put plot at right:

   CIRCLE: increasing or decreasing

   CIRCLE: concave up or concave down

   (c) $A = \int_1^5 f(x)\,dx =$ ______________________________

2. Left endpoint Riemann sum:

   (a) Put plot at right:

   (b) CIRCLE: upper bound or lower bound

   Reason:

3. Right endpoint Riemann sum:

   (a) Put plot at right:

   (b) CIRCLE: upper bound or lower bound

   Reason:

4. Trapezoid rule:

   (a) Reason

   (b) Put plot at right:

   (c) CIRCLE: upper bound or lower bound

       Reason:

5. Midpoint Riemann sum:

   (a) Reason:

   (b) Put plot at right:

   (c) CIRCLE: upper bound or lower bound

       Reason:

6. Simpson's rule:

   (a) *20% extra credit.* Use your own paper.

7. *20% extra credit.* Error Analysis. Use your own paper.

Approximation to $\int_1^5 f(x)\,dx$:

n	Δx	Left sum	Right sum	Trapezoid	Midpoint	Simpson
4	1					
16	1/4					
64	1/16					
256	1/64					

Error = Approximation $- \int_1^5 f(x)\,dx$ **:**

n	Δx	Left sum	Right sum	Trapezoid	Midpoint	Simpson
4	1					
16	1/4					
64	1/16					
256	1/64					

Shifting and Rescaling Functions (Two Drills)

Objectives: The purpose of these drills is to help you understand the operations on a function that shifts or rescales its graph.

Prerequisites: You should be able to identify the graphs of the following seven basic functions

$$x, \qquad \frac{1}{x}, \qquad x^2, \qquad x^3, \qquad \sqrt{x}, \qquad x^{1/3}, \qquad |x|$$

You should also understand the following concepts. (If you are uncertain about them, they will become clear as you work through the drills.)

If you start with the graph of a function $f(x)$, then the graph of the function $f(x)+b$ will be shifted up or down by b, whereas the graph of the function $f(x-a)$ will be shifted right or left by a.

If you start with the graph of a function $f(x)$, then the graph of the function $bf(x)$ will be expanded or contracted vertically by a factor $|b|$ and will be reflected (vertically) through the x-axis if b is negative. Further, the graph of the function $f(x/a)$ will be expanded or contracted horizontally by a factor $|a|$ and will be reflected (horizontally) through the y-axis if a is negative.

***Mathematica* Commands:** You are not expected to know any *Mathematica* commands except how to enter an expression and how to execute a command.

Initialization:

1. Open the *Mathematica* Notebook called `shift.ma` by double-clicking on its icon. The Notebook `shift.ma` is in the directory `chap15/shift`. Your instructor will tell you how to find this file.
2. Because the plots will appear in the Notebook and your answers will be given in a separate pop-up window, you may wish to reposition the answer window so that it does not obscure the Notebook window.

Procedure:

1. There are two separate drills in this *Mathematica* Notebook. One is on shifting functions and the other is on rescaling functions.

2. You need to execute start commands for both drills: one is called `StartShiftDrill[]` and the other is called `StartScaleDrill[]`. To execute either of these commands, position the cursor on the line and press ⟨SHIFT⟩ and ⟨RETURN⟩ simultaneously. You will be shown the graph of a function. Your goal is to enter the formula for this function in the answer window that will appear once you start either drill.

3. For the drill on shifting functions, the graph is one of the seven basic functions shifted up or down and left or right. For example, if the basic function is x^2, the graph is $(x - a)^2 + b$ for some integers a and b. From the graph, you identify the basic function and the numbers a and b. Then enter the formula on the answer line and execute it.

4. For the drill on rescaling functions, the graph is one of the seven basic functions expanded or contracted and possibly reflected vertically and expanded or contracted and possibly reflected horizontally. For example, if the basic function is x^2, the graph is $b(x/a)^2$ for some integers a and b. From the graph, you identify the basic function and the numbers a and b. Then enter the formula on the answer line and execute it.

5. After you have executed the answer line, *Mathematica* will check your answer. If you make a mistake, you may change your answer and recheck it any number of times. If you are lost, you can type “show me” in the answer window to see the answer.

6. When you are finished with a given graph, delete all the plots and then reexecute the start command for the appropriate drill to get a new graph.

7. You may jump back and forth between the two drills as often as you like.

Lab Report Requirements: If you do this drill as a lab, run through the drill as many times as necessary until you feel confident in your ability. You should expect to have a quiz or a test question on the subject. A question might show you the graph of a function and ask you to write down its formula, or it might give you a formula and ask you to draw its graph.

16 Projects

The Flight of a Baseball

Objectives: Imagine that a baseball player is at home plate and hits the ball in the air. What parametric equations describe the position of the ball t seconds after it is hit? How far will the ball travel? How fast does the ball need to be hit to clear the home run fence? The following discussion and exercises are designed to answer these questions. Use *Mathematica* to help you with the (extensive) computations and graphics.

Background: First, consider a simplified model that ignores air resistance. In this case, after the ball is hit, the only force acting on the ball is the vertical force caused by gravity. Therefore, the following equations describe the x- and y- components of the acceleration of the ball

$$\frac{d^2x}{dt^2} = 0 \quad \text{and} \quad \frac{d^2y}{dt^2} = -g$$

where g is the acceleration constant caused by gravity ($g = 32$, where the unit of distance is feet). Integrating these equations with respect to t gives

$$\frac{dx}{dt} = C_1 \quad \text{and} \quad \frac{dy}{dt} = -gt + C_2$$

where the constants C_1 and C_2 can be evaluated by considering the initial velocity of the ball. Suppose the initial speed of the ball is the constant v (in units of feet per second) and suppose the angle of inclination of the ball is A. Then, the x- and y- components of the initial velocity of the ball are given by $v\cos(A)$ and $v\sin(A)$, respectively. After substituting $t = 0$ in the above equations and solving for C_1 and C_2, we obtain the following equations:

$$\frac{dx}{dt} = v\cos(A) \quad \text{and} \quad \frac{dy}{dt} = -gt + v\sin(A)$$

One more integration with respect to t yields

$$x = vt\cos(A) + c_1 \quad \text{and} \quad y = \frac{-gt^2}{2} + vt\sin(A) + c_2$$

Assume that the origin ($x = 0, y = 0$) is the location of home plate and that the shoulder height of the batter (from where the ball is hit) is h feet. Then, by substituting $t = 0$, we can find the constants c_1 and c_2. The final parameterization of the baseball is given by

$$x = vt\cos(A) \quad \text{and} \quad y = \frac{-gt^2}{2} + vt\sin(A) + h$$

Instructions: Solve the following problem, which has 4 parts.

1. Set $g = 32$, the angle A at $\pi/4$, the velocity v to be 120 feet per second, and the height h to be 6 feet. Define the x- and y- coordinates of the ball t seconds later (see the preceding equations for x and y) as functions of t and then plot the trajectory of the ball until the ball hits level ground. Start the parameter t at 0, and experiment with different terminal values of t to try and get the entire flight of the baseball (until the ball hits the ground) on the screen. Find the horizontal distance traveled by the ball and the elapsed time.

   Recall that, to plot a parameterized curve (x, y), where x and y are expressions of t, $a \leq t \leq b$, we enter the command

   ```
   ParametricPlot[{x, y}, {t, a, b}]
   ```

2. What is the shape of the graph? Eliminate t, and determine the equation of the trajectory in the form of y as a function of x. To do this, solve for t in terms of x (this you can do by hand), and then use the `ReplaceAll` command to substitute this expression for t into y.

   Now make the angle A and the initial speed v and t free variables `Clear[A, v, t]` and reinput the equations for x and y into *Mathematica.*

3. Suppose the home run fence is 10 feet high and 350 feet from home plate. What is the minimum velocity at which the ball must leave the bat so that the ball just clears the home run fence? *Be careful:* Do not assume any particular value of the angle A. In fact, your strategy should be as follows. First, eliminate t as you did in part 2 (only now using v and A as free variables). Then solve for the speed v in terms of the angle A, so that the parameterized equations for x and y pass through the point $x = 350$, $y = 10$. Finally, minimize v as a function of t.

4. *Extra credit. Research project.* Now assume air resistance acts on the ball. Air resistance acts in the opposite direction to the velocity of the ball and its magnitude is proportional to the speed. This leads to the acceleration equations

   $$\frac{d^2x}{dt^2} = -k\frac{dx}{dt} \quad \text{and} \quad \frac{d^2y}{dt^2} = -g - k\frac{dy}{dt}.$$

   Here, k is a friction constant, which will be given later. Solve these equations for x and y. Take the limit of your solutions as $k \to 0$ and see if your result agrees with the solution for x and y without air resistance. Then, repeat parts 1 and 3, taking into account air resistance with $k = 0.1$. Compare your plots and your answers with your results that did not consider air resistance. For part 3, you will not be able to algebraically solve for v in terms of A (even with *Mathematica*). Instead, take specific values of A (near $\pi/4$) and solve for v numerically (using `FindRoot`). Determine an approximate minimum value for v.

Curves Generated by Rolling Circles

Objectives: The goal of this project is to use *Mathematica*'s plotting and integration commands to help answer questions on the *cycloid*, which is a curve generated by a point on a rolling circle. A more complicated version of the cycloid is also considered.

Background: Consider a wheel of radius R. Fix a point on the rim of the wheel. Now let the wheel roll on level ground and consider the path traced out by the point P (see the figure). This path is called the cycloid.

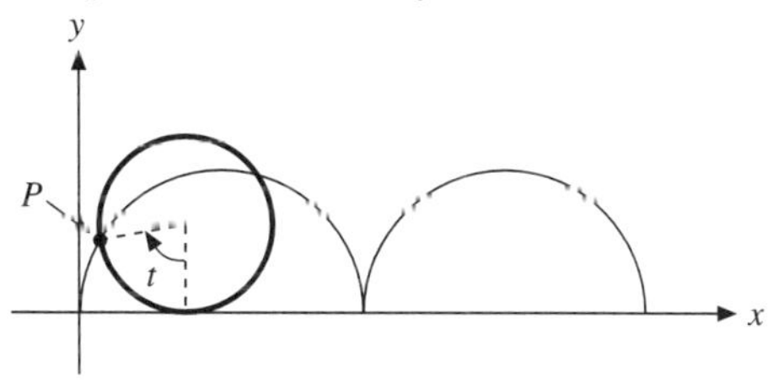

Instructions: Answer the following questions, using *Mathematica* where appropriate.

1. Show that the cycloid is parameterized by the formulas

$$x(t) = R(t - \sin(t)) \text{ and } y(t) = R(1 - \cos(t))$$

Here, t is the angle between the vertical and the ray that extends from the center of the circle to P (so $t = 0$ when P is at the origin).

2. Use *Mathematica* to plot two arches of this cycloid with $R = 1$. To plot a parametric curve (x, y) (where x and y are expressions in t) over the interval $a \leq t \leq b$, use the command

```
ParametricPlot[{x, y}, {t, a, b}, AspectRatio->Automatic]
```

Note the square brackets that are used with this plot command. Now unassign R (`Clear[R]`).

3. Compute the arc length of one arch of this cycloid (for a general value of R). Recall that the arc length of a parameterized curve $(x(t), y(t))$ for $a \leq t \leq b$ is given by

$$\int_a^b \sqrt{(x')^2 + (y')^2}\, dt$$

4. In the previous problem, you determined the arc length of one arch by computing the integral over the interval $0 \le t \le 2\pi$. Presumably, computing the appropriate integral over the interval $0 \le t \le 4\pi$ should give the arc length of two arches. Check this hypothesis, and explain why you think *Mathematica* gives a warning about the convergence of the answer it returns. Should you be concerned?

5. The slope of a parameterized curve $(x(t), y(t))$ is given by

$$\frac{dy}{dx} = \frac{dy/dt}{dx/dt}$$

   Use *Mathematica* to show that the limit of the slope of the tangent line as $t \to 0^+$ is infinity.

   Note: It may be easier to show that 1/slope $\to 0$ as $t \to 0^+$.

6. Now suppose that a circle of radius a rolls around the outside of the circle of radius $R > a$ centered at the origin (see the accompanying diagram). Find the parameterization $P = (x(t), y(t))$ that describes the path of a fixed point P on the rolling circle. Here t is the angle measured counterclockwise from the positive x-axis to the line segment that runs from the origin to the center of the rolling circle. Assume that P is located at the point $(R, 0)$ when $t = 0$. Compute the arc length of one of the arches of this path.

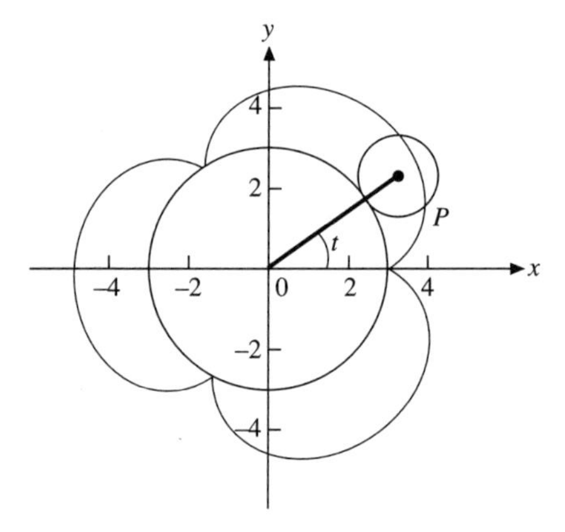

The Duck Hunt

Suppose that you are sitting in a duck blind, which we will assume is located at the origin. Assume that this is a two-dimensional problem. At time zero, using your binoculars, you spot a duck flying in your direction. From past experience with ducks, you know that it will be flying along the curve

$$y = f(x) = 21 + \sin(x) + 1/20x^2$$

where x is in feet and denotes the horizontal distance from the duck blind. You also know that the speed of the duck is 45 miles per hour, and that its maximum speed is 80 miles per hour. For those of us who are not duck experts, these numbers are attainable by canvasback ducks. When you first see the duck at time zero, it is located at $(-150, f(-150))$. The velocity of the shot in your gun is 600 feet per second, and the effective range is 120 feet. You also know that, as soon as you fire that first shot, the duck will change course and fly upward in a direction perpendicular to the direction from which the shot was fired (assuming, of course, that you miss), with an acceleration of 10 feet per second per second until it attains its maximum speed of 80 miles per hour.

Answer the following questions.

1. When is the duck in range?
2. When is the duck closest to you?
3. If you wish to shoot the duck when it is directly overhead, when do you pull the trigger?
4. Assume that you take a shot when the duck is closest to you and that you miss. If it takes you one second to fire a second shot, will the duck still be in range when you fire? Explain.

Gravitational Force

Objectives: The goal of this project is to determine the gravitational force between a mass m that is concentrated at a point and other objects whose mass may be spread out over a large region—for example, a ring, a spherical shell, or a solid ball (such as the Earth).

Background: This project assumes knowledge of integration and the area of a surface of revolution. The gravitational attraction between two bodies of mass m and M (which are concentrated at two points in space) is given by the following inverse square formula

$$G\frac{mM}{r^2}$$

where r is the distance between the two masses and G is the gravitational constant (whose value depends on the units involved). If one of the objects, say M, is spread out over a large region (such as the Earth), then the distance between m and various points of M will vary (i.e., there is no well-defined value for r). To determine the gravitational attraction between m and a large object M, the large object must be subdivided into smaller pieces whose gravitational attraction with m is easy to compute. Then, the gravitational attractions of the smaller pieces must be summed (integrated) to determine the total gravitational attraction. *Mathematica* can help with the computations.

Instructions: Complete the following problem that will lead to the computation of the gravitational force between a point mass m and a large homogeneous spherical ball M. Use complete sentences to describe your solution.

1. *Surface area property of the sphere* (this result will be used later). Consider the sphere of radius R centered at the origin. Slice this sphere along two parallel planes. Show that the surface area of the slice depends only on the distance between the two planes (and not on their locations). *Hint:* Think of the sphere as the surface obtained by revolving a circle of radius R, centered at the origin, about the x-axis. Let the planes be perpendicular to the x-axis at $x = a$ and $x = b$, where $-R \leq a \leq b \leq R$. Now show that the surface area of revolution of this circle between $x = a$ and $y = b$ depends only on $b - a$. This problem can be done by hand or by using *Mathematica*.

2. Show that the magnitude of the gravitational attraction between the point mass m and a uniform circular wire of radius r and total mass M, whose center is a distance h from m, is

$$\frac{GmMh}{(h^2 + r^2)^{\frac{3}{2}}}$$

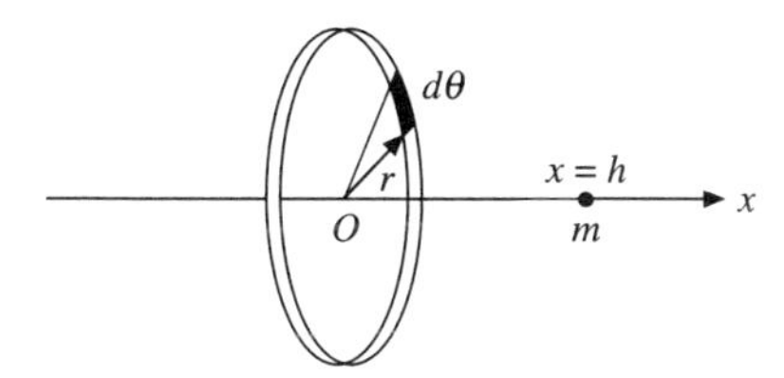

Hint: Divide the circular wire of mass M into angular sectors of angular width $d\theta$. Each sector has mass $M d\theta/(2\pi)$. By symmetry, only the component of the gravitational force that is parallel to the x-axis needs to be computed (because the component perpendicular to the x-axis will cancel with the perpendicular component of the analogous angular sector on the opposite side of the circle). Then, add up (integrate) over $0 \leq \theta \leq 2\pi$. This integral is simple enough to do without *Mathematica*.

3. Show that the gravitational force between the point mass m and a hollow spherical shell of radius R and mass M is

$$\frac{GmM}{a^2}$$

where $a > R$ is the distance between the mass m and the center of the shell.

This means that the gravitational attraction is the same as if all the mass of the spherical shell were concentrated at the center.

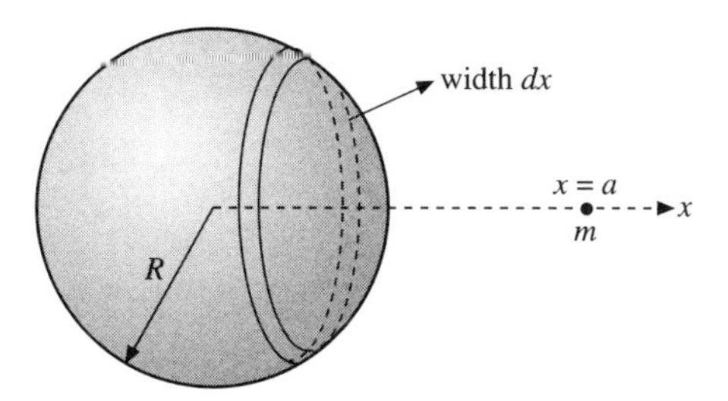

Hint: Divide the shell into rings of width dx that are perpendicular to the x-axis. Let x be the coordinate that represents the center of the typical ring (so x can range from $-R$ to R). In view of part 1, the mass of this ring is independent of x (it just depends on the width, dx). Use this result to show that the mass of the ring is $Mdx/2R$, and then use part 2 to show that the gravitational attraction between the mass m and this ring is

$$\frac{GmM}{2R}\left(\frac{a-x}{(a^2+R^2-2ax)^{\frac{3}{2}}}\right)dx$$

Now, with the help of *Mathematica*, compute the gravitational attraction between m and the spherical shell, by integrating this expression from $x=-R$ to $x=R$. You will need to simplify your answer. Note that $a>0$, $R>0$ and $a>R$.

4. In part 3, the mass m is assumed to be outside the shell. It is known that if the mass is inside the shell, (i.e., if $a<R$), then the gravitational attraction between the mass m and the spherical shell is zero. Use *Mathematica* to verify this axiom by reexecuting the above integral but, this time, realizing $a<R$.

5. Show that the gravitational attraction between a point mass m and a uniform ball of radius R and mass M is

$$\frac{GmM}{a^2}$$

where a is the distance between the mass m and the center of the ball. *This means that the gravitational attraction between the mass m and the ball is the same as if all the mass were concentrated at the center of the ball*—just as in the case of a spherical shell.

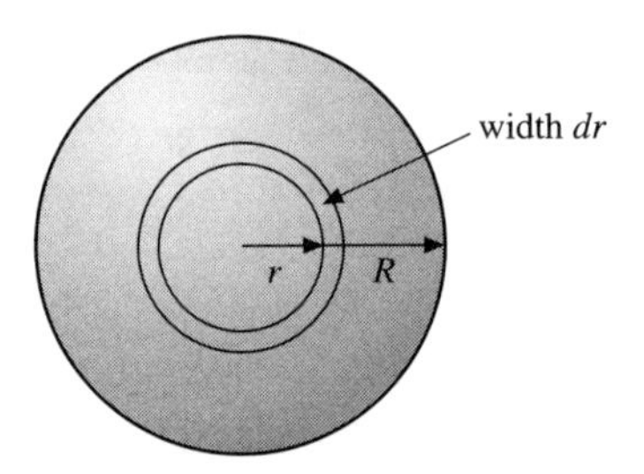

Hint: Let λ denote the (constant) mass density of the ball. Divide the ball into spherical shells of radius r with thickness dr. The mass of this shell is $4\pi\lambda r^2 dr$—i.e., (mass density) $\times$ (the volume of the shell). Use part 3 to find the gravitational attraction between the mass m and this spherical shell. Then integrate over r from $r = 0$ to $r = R$ and use the fact that $\lambda \times$ (the volume of the ball) $= M$. (This integral is simple enough that *Mathematica* is not required.)

6. For a very large mass such as the Earth, it is more realistic to treat the density λ as a function of r, the distance to the center of the ball. Show that the answer to the previous problem is unchanged in this case. *Hint:* First express the mass M of the ball as an integral in terms of $\lambda = \lambda(r)$. You do not have to know what $\lambda(r)$ is!

Logistic Growth

The accompanying table contains population data for the United States. The logistic equation is a first-order nonlinear differential equation of the form

$$\frac{dP}{dt} = aP - bP^2$$

where P in this example represents the population of the United States at time t, in years. By picking an appropriate starting value for $P(0)$, find those constants a and b such that the solution to the differential equation best fits the data in the table.

Year	Population	Year	Population
1790	3.93	1900	75.99
1800	5.3331	1910	91.97
1810	7.24	1920	105.71
1820	9.64	1930	122.78
1830	12.87	1940	131.67
1840	17.07	1950	151.333
1850	23.19	1960	179.32
1860	31.44	1970	203.21
1870	39.82	1980	226.5
1880	50.16	1990	249.63
1890	62.95	2000	?

Note that the population is given in millions. Thus, the population of the United States in 1790 is estimated to be 3,930,000.

By estimating the slope of the curve at various points, you should be able to get estimates for the unknown parameters a, b, and $P(0)$. Plot $P(t)$ for various choices of the parameters and choose those parameters that give a curve that best fits the data. Estimate the population of the United States in the year 2000. Do you think this is a reasonable model for predicting the population in the year 2000? Can you think of a better one?

Search for the Meteor

Objectives: The goal of this project is to solve the following problem. A meteor crashes somewhere in the hills that lie north of point A. The impact is heard at point A and, 6 seconds later, it is heard at point B. Three seconds still later, it is heard at point C. Locate the point of impact of the meteor, given that A lies four miles due east of B and two miles due west of C. The speed of sound is roughly .20 miles per second.

Instructions: Solve the given problem. Use complete sentences to describe your procedure. Include an accurate plot with your explanation.

Background and Hints: Read about hyperbolas in your calculus textbook. There you will find that the equation of a hyperbola, centered at the origin and with focal points at $+c$ and $-c$ along the x-axis, is

$$\frac{x^2}{a^2} - \frac{y^2}{c^2 - a^2} = 1$$

Here, $2a$ represents the difference $|d_1 - d_2|$, where d_1 is the distance between an arbitrary point (x, y) on the hyperbola and the focal point $+c$ and d_2 is the distance between (x, y) and the other focal point $-c$.

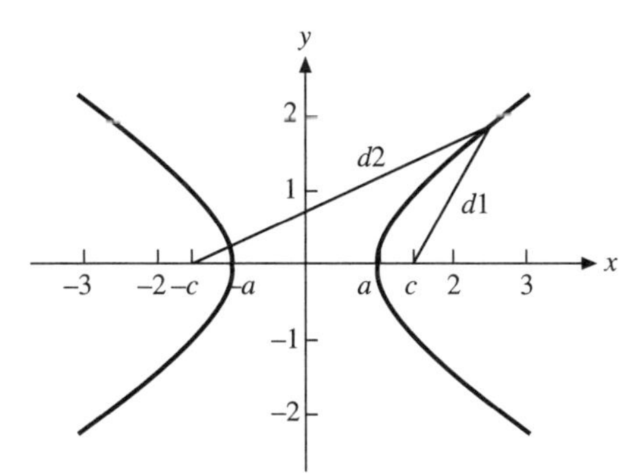

For this project, locate points A, B, and C on the x-axis. The choice of origin is somewhat arbitrary, but a convenient choice of origin is the midpoint between points A and B. From the information given, the meteor must lie on a hyperbola centered at the origin with points A and B as focal points. Find the equation of this hyperbola. Likewise, the meteor must also lie on a hyperbola with focal points at B and C. (Where is this hyperbola centered?) Thus, the meteor must lie on the point of intersection between the two hyperbolas.

Note: A hyperbola has two branches and you must take care to choose the correct branch. The information given in the problem should lead you to the correct branch.

Mathematica will be useful in solving the two equations that represent the two hyperbolas. To obtain a plot of the hyperbolas, use the `ImplicitPlot` command (see Chapter 2).

Radioactive Waste at a Nuclear Power Plant

A nuclear power plant produces a waste product that is a radioactive isotope, called A. The isotope A has a half-life of 10 years; 70% by weight decays into a radioactive isotope B, and 30% into a radioactive isotope C. The isotope B has a half-life of 20 years and decays into nonradioactive byproducts. The radioactive isotope C has a half-life of 30 years and also decays into nonradioactive byproducts. Answer the following questions.

1. Suppose you start with 400 kilograms of isotope A. What is the maximum amount of isotope B that will be present, and when will this occur? Answer the same question for isotope C.

2. When the power plant began operating, there was no isotope A, B, or C present. If the power plant operates so that it produces isotope A at the constant rate of 40 kilograms per year, what are the maximum amounts of isotopes A, B, and C that will be present, and when will these different maxima occur?

3. Federal safety requirements specify that the reactor can never have on hand more than 500 kilograms of isotope A, 400 kilograms of isotope B, or 300 kilograms of isotope C. What is the maximum rate at which the power plant can produce isotope A without violating the federal regulations?

You may assume that isotopes B and C weigh essentially the same as isotope A. Thus, for example, if 100 kilograms of A decays there will be 70 kilograms of B and 30 kilograms of C.

Let $A(t)$, $B(t)$, and $C(t)$ denote the amounts of isotopes A, B, and C that are present at time t.

Hints:

(a) First find the three decay constants.

(b) If a radioactive isotope is being produced by some source at the same time as it is decaying, how does that alter the differential equation for the rate of change of the amount of this isotope?

(c) Be sure to plot A, B, and C, to ascertain if they have the qualitative behavior you expect.

Pension Funds

Background: A pension fund starts out with $\$P$ (at $t = 0$) and is invested with a return of $100r\%$ per year, compounded continuously (here, r is the interest rate, given as a number between 0 and 1). The pension fund must continuously pay out money at the rate of $\$R$ per year to its employees for a period of n years (this means that the value of the pension fund decreases to zero after n years). Let $y(t)$ denote the value of the pension fund after t years.

Instructions: Answer the following questions with complete sentences.

1. From the information given, derive the differential equation $y' = ry - R$, with the conditions $y(0) = P$, $y(n) = 0$.
2. Solve this differential equation for y.
3. Find a formula for P in terms of r, n, and R (P represents the amount of money required to pay out $\$R$ per year for n years, assuming the rate of return on the investment is $100r\%$).
4. Calculate P for $R = 50K$, $r = .07$, and $n = 20$.
5. Calculate the interest rate (r) required so that an initial value of $P = 500K$ for the pension fund will pay out $50K$ per year for 30 years.

***Mathematica* Hints:** *Mathematica* is not needed in an essential way for parts 1-4. However, you can check your solution to the differential equation (part 2) by using *Mathematica*'s `DSolve` command for solving differential equations. To use this command, first enter the differential equation and give it a label (such as *eq*).

```
eq = (D[y[t],t] == r y[t] - R);
```

Then, to solve this differential equation, issue the command

```
DSolve[eq, y[t], t]
```

The general solution to the differential equation is returned, with an unknown constant of integration. This constant can be evaluated using the initial condition $y(0) = P$. The differential equation with the initial condition can be solved with the command

```
DSolve[{eq, y[0]==P}, y[t], t]
```

In part 5, *Mathematica* is necessary to solve the relevant equation for r.

The Center of the State of Texas

Objectives: The goal of this project is to compute (an approximation to) the center of mass of the state of Texas from data (given here) that represent the state's boundary.

Background: The center of mass of a region bounded above by $y = f(x)$ and below by $y = g(x)$, $a \leq x \leq b$, is the point $(x0, y0)$ where

$$\mathrm{x0} = \frac{1}{A}\int_a^b x[f(x) - g(x)]\, dx$$

$$\mathrm{y0} = \frac{1}{A}\int_a^b 1/2\left([f(x)]^2 - [g(x)]^2\right)\, dx$$

and A represents the area of the region. For the state of Texas, the origin will be located at the western tip (near El Paso) and the x-axis is the extension of the east-west border between New Mexico and Texas. In the preceding integral formulas, the graph of $y = f(x)$ represents the upper boundary and the graph of $y = g(x)$ represents the lower boundary of the region in question. Because there are no analytical formulas for f and g for the boundary of the state of Texas, approximations to the given integrals must be computed using the following data, which represent the boundary of the state. The second coordinate represents the values of f and g at integer values of x from $x = 0$ to $x = 11$, where each unit represents 69 miles.

```
north={{0,0},{1,0},{2,0},{3,0},{3,4.5},{4,4.5},{5,4.5},
       {6,4.5},{6,2.2},{7,2.1},{8,1.8}, {9,1.9},{10,1.8},
       {11,1.7},{11,-2.2}};
south={{0,0},{1,-1.1},{2,-2.5},{3,-2.9},{4,-2.3},{5,-2.8},
       {6,-4.4},{7,-5.8},{8,-6.1},{9,-3.3},{10,-2.8},
       {11,-2.2}};
```

Note that there are two y-values given in the northern boundary for both $x = 3$ and $x = 6$ (because $x = 3$ and $x = 6$ represent the two north-south boundaries of the Panhandle).

Instructions: Use these data to compute the area A of the state of Texas using trapezoidal rule. Then, determine the center of mass of the state using the trapezoidal rule to compute the integrals. Include a plot that shows the boundary of the state with its center of mass.

***Mathematica* Hints:** Enter the preceding data as lists labeled *north* and *south* (or make up your own names). To refer to the entries in a list, use brackets [[]]; for example, `south[[3]]` refers to the point $[2, -2.5]$ and `south[[3,2]]` refers to the second entry of this point (i.e., -2.5). For example, the following command sumsthe second entries (the y-values) of all the points on this list.

```
Sum[ south[[i, 2]], {i, 1, 12}]
```

To plot the boundary of the state, enter the command

```
Show[ListPlot[north,
              PlotJoined->True,
              AspectRatio->Automatic,
              DisplayFunction->Identity],
     ListPlot[south,
              PlotJoined->True,
              AspectRatio->Automatic,
              DisplayFunction->Identity],
     DisplayFunction->$DisplayFunction]
```

You can set `AspectRatio->Automatic` to make a plot with the same scale on the horizontal and vertical axes. Also, you can set `DisplayFunction->Identity` and later `DisplayFunction->$DisplayFunction` to control exactly when the plot is drawn on the screen.

To plot the state boundary along with the center point $[x0, y0]$, issue the following commands

```
texas = Show[ListPlot[north,
                      PlotJoined->True,
                      AspectRatio->Automatic,
                      DisplayFunction->Identity],
             ListPlot[south,
                      PlotJoined->True,
                      AspectRatio->Automatic,
                      DisplayFunction->Identity]];
center = Graphics[Point[{x0,y0}]];
Show[texas, center,
    DisplayFunction->$DisplayFunction]
```

The first two commands assign the labels *texas* and *center* to the state boundary and the point $[x0, y0]$, respectively. The third line of commands displays both plots on the same plot window.

The Brightest Phase of Venus

Objectives and Background: This project requires knowledge of the material on max/min and area—Chapters 6 and 8. The brightness of Venus is proportional to the area of the visible portion of Venus and inversely proportional to the square of the distance from the Earth to Venus. From the accompanying figure, note that, as the angle t increases from 0 to π, the area of the visible portion of Venus increases. This phenomenon tends to increase the brightness of Venus. But the distance d from the Earth to Venus also increases, which tends to decrease the brightness of Venus. For some angle t, between 0 and π, Venus will appear brightest. Find this angle.

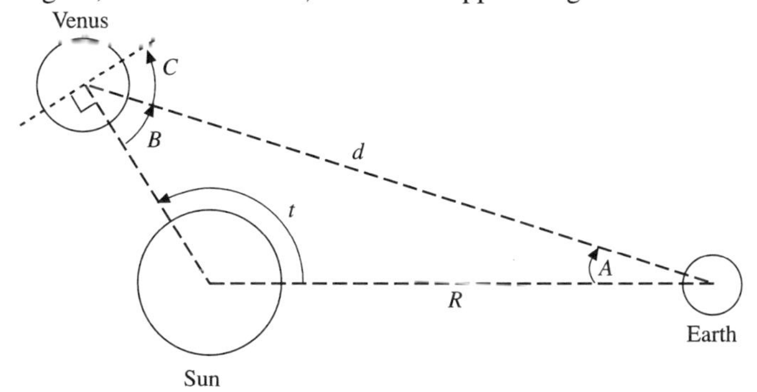

Instructions: Find the brightest phase of Venus (i.e., the angle t) by following the outline given here. Write up your solution using complete sentences. Use *Mathematica* where appropriate to help with the computations and graphics.

1. As mentioned above, brightness is proportional to the quantity

$$\frac{\text{Area of visible portion of Venus}}{d^2}$$

2. From geometry, show that $B = \pi - (t + A)$ and $C = (t + A) - \pi/2$.

3. Let a be the radius of Venus. Show that the visible portion of Venus (from Earth) lies between the curves $x = -\sqrt{a^2 - y^2}$ and $x = \sin(C)\sqrt{a^2 - y^2}$. Now show that the area of the visible portion of Venus is

$$\frac{\pi a^2}{2}(1 + \sin(C))$$

4. Combine parts 1, 2, and 3 to obtain a formula for the brightness of Venus that depends on the angles t and A and the distance d. The goal is to express the brightness in terms of one variable t. To this end, use the law of cosines and the law of sines to show the following

$$d^2 = r^2 + R^2 - 2rR\cos(t)$$

$$\sin(A) = \frac{r\sin(t)}{d}$$

Here, r is the distance from the Sun to Venus and R is the distance from the Sun to the Earth. Use the values $R = 93$, $r = 67$, and $a = 0.004$ (the unit of distance is 1 million miles).

5. Use the equations in part 4 to find an expression for the brightness of Venus that depends on one variable t. Use *Mathematica* to maximize this function over the interval $0 \leq t \leq \pi$.

Index

Enhance your understanding of calculus with the problem-solving power of *Mathematica*®.

This introductory text/lab manual/resource guide puts the focus on using the Mathematica computer algebra system to increase your conceptual knowledge of calculus and improve your problem-solving skills. Follow ***CalcLabs With Mathematica***'s easy-to-use presentation, and solve the problems in your calculus course more quickly and easily.

In Part One you'll learn how to:

- find maxima and minima of functions
- calculate limits
- locate inflection points and asymptotes
- solve differential equations
- integrate and differentiate expressions
- visualize integration techniques
- animate sequences of Taylor approximations
- program Mathematica to do calculus
- and much more!

With Part Two's projects and weekly labs, you'll learn how to use Mathematica appropriately and master its more sophisticated capabilities. You'll discover that Mathematica is an indispensable problem-solving tool for your calculus needs.

Additional Mathematica-related from Brooks/Cole:

Linear Algebra With Mathem
Eugene Johnson

Tutorial Introduction to Mathem
Wade Ellis/Ed Lodi

ISBN 0-534-34086-5